GeoPlanet: Earth and Planetary Sciences

For further volumes:
http://www.springer.com/series/8821

Waldemar Walczowski

Atlantic Water in the Nordic Seas

Properties, Variability, Climatic Importance

Springer

Waldemar Walczowski
Institute of Oceanology,
Physical Oceanography
Polish Academy of Sciences
Sopot, Pomorskie
Poland

The GeoPlanet: Earth and Planetary Sciences Book Series is in part a continuation of Monographic Volumes of Publications of the Institute of Geophysics, Polish Academy of Sciences, the journal published since 1962 (http://pub.igf.edu.pl/index.php).

ISSN 2190-5193 ISSN 2190-5207 (electronic)
ISBN 978-3-319-01278-0 ISBN 978-3-319-01279-7 (eBook)
DOI 10.1007/978-3-319-01279-7
Springer Cham Heidelberg New York Dordrecht London

Library of Congress Control Number: 2013944781

Printed on acid-free paper

Springer is part of Springer Science+Business Media (www.springer.com)

Series Editors

Managing Editor

Advisory Editor

Kon-Kee Liu
Institute of Hydrological
and Oceanic Sciences
National Central University Jhongli
Jhongli, Taiwan

Teresa Madeyska
Research Centre in Warsaw
Institute of Geological Sciences
Warszawa, Poland

Stanisław Massel
Institute of Oceanology
Polish Academy of Sciences
Sopot, Polska

Antonio Meloni
Instituto Nazionale di Geofisica
Rome, Italy

Evangelos Papathanassiou
Hellenic Centre for Marine Research
Anavissos, Greece

Kaja Pietsch
AGH University of Science and
Technology
Kraków, Poland

Dušan Plašienka
Prírodovedecká fakulta, UK
Univerzita Komenského
Bratislava, Slovakia

Barbara Popielawska
Space Research Centre
Polish Academy of Sciences
Warszawa, Poland

Tilman Spohn
Deutsches Zentrum für Luft-
und Raumfahrt
in der Helmholtz Gemeinschaft
Institut für Planetenforschung
Berlin, Germany

Krzysztof Stasiewicz
Swedish Institute of Space Physics
Uppsala, Sweden

Roman Teisseyre
Earth's Interior Dynamics Lab
Institute of Geophysics
Polish Academy of Sciences
Warszawa, Poland

Jacek Tronczynski
Laboratory of Biogeochemistry
of Organic Contaminants
IFREMER DCN_BE
Nantes, France

Steve Wallis
School of the Built Environment
Heriot-Watt University
Riccarton, Edinburgh
Scotland, UK

Wacław M. Zuberek
Department of Applied Geology
University of Silesia
Sosnowiec, Poland

It is evident that oceanographic conditions of the North Polar Basin have much influence on climate, and it is equally evident that changes in the conditions of circulation would greatly change the climatic conditions.

Fridtjof Nansen (1902)
(Polar Research, 2001, 20(2), p. 127)

Preface

Climatic changes observed in recent years are worrying. A few years ago it was not completely obvious whether the climate was actually getting warmer, or was just a temporary fluctuation. Today, we know that these are not just typical weather anomalies, as since 2001 each subsequent year has been almost the warmest year in the history of world meteorological measurements. We have evidence that it is getting warmer in the global scale and we—humans—are responsible for that. The Fourth Assessment Report of the Intergovernmental Panel on Climate Change (IPCC) leaves very small margin of error. The probability that the changes result from the greenhouse gas emission exceeds 90 % (IPCC AR4, 2007). While the direct causes of climate warming have been recognized, the mechanisms and long-term effects of the processes remain unknown or not completely clear.

The Earth's climate turned out to be an extraordinarily complex phenomenon, affected by a great deal of factors and processes. One of the basic climate control factors, yet underestimated until recently, is the ocean. The significance of the sea for mitigating climate has been always recognized—at the seaside winters are warmer and summers cooler. The role of the Gulf Stream in warming the Atlantic part of Europe was known.

However, the ocean had not been treated as a primary global-scale climate control factor, affecting our planet in a very wide range of processes and various timescales. Only the recent decades, particularly the recent years of increased interest in the climate, has ignited a revolution, an upheaval in thinking about the ocean's role, the quantity of heat it receives, stores and transports, and the potential consequences of hypothetical change in the ocean circulation pattern. It was found that the ocean affects climate not only as an important ingredient of the planetary hydrographic and energy cycle, but also participates in the biogeochemical cycle, exchange of gas with the atmosphere which, in turn, affects the greenhouse effect. It is, for instance, a great reservoir, a buffer which stores the most common and the most dangerous greenhouse gas released by human civilization—carbon dioxide.

In spite of the increasing recognition of the ocean's role, even today, the most renowned report concerning the climate (AR4, IPCC, 2007) contains the following sentence: "Although scientists now better appreciate the strength and variability of the global-scale ocean circulation, its roles in climate are still hotly debated. Is it a passive recipient of atmospheric forcing and so merely a diagnostic consequence of climate change, or is it an active contributor?" Ocean researchers have no doubts: it is an extremely important and active climate control factor. The observed changes influence the ocean and the ocean actively responds to them and modifies them. The Atlantic Arctic has a special role in the processes.

Sopot, July 2008 Waldemar Walczowski

Credits and Acknowledgments

Besides the main purpose of the present book, which was to describe and diagnose the area of the Nordic Seas studied by the Institute of Oceanology of the Polish Academy of Sciences from the point of view of the physical oceanography, the author's intention was also to place it in a certain context. The context of people, equipment, development of knowledge and measuring technology, and the evolution of ideas.

The author took part in establishing the Ocean Circulation Laboratory (previously known as the Marine Hydrography and Physical Oceanography Laboratory), under the management of Prof. Jan Piechura, the head of the Physical Oceanography Department. Together with Agnieszka Beszczyńska-Moeller, Robert Osiński, Piotr Wieczorek and later with Ilona Goszczko, Małgorzata Kitowska, Jaromir Jakacki, and Daniel Rak, we have prepared research programs, gathered new equipment, and learnt how to use it, developed the required software and provided equipment for the ship. The results occurred after years of work—we became a partner of international programs, we participate in conference, we publish in worldly recognized periodicals. We have gained what is most important—knowledge about the ocean. The physical type of it, but also the type that cannot be found in books or databases, but only through hundreds of hours spent working at sea. Science is also an evolution of ideas, creating new concepts, and collecting experience.

I wish to thank those who made it possible for me to work on the sea—colleagues from the laboratory as well as Longin Stojek, Jerzy Dąbrowski, the ship crew, including in particular Roman Obuchowski, Krzysztof Rosiński, seamen, and mechanicians.

Some of the colleagues—Asia Lech, Zenon Płachta—have already passed away, yet they are still in our memory. I would also like to thank those who supported me in my work on the paper: Agnieszka Beszczyńska-Moeller, Marek Zwierz, and Tomasz Petelski. I would like to express my deep gratitude to reviewers of the thesis—Prof. Jan Piechura and Prof. Jerzy Cyberski—for great patience and assistance.

Contents

Abbreviations

ACIA	Arctic Climate Impact Assessment
AD	Atlantic Domain
ADCP	Acoustic Doppler Current Profiler
ADU	Ashtech Attitude-sensing System
AF	Arctic Front
AMO	Atlantic Multidecadal Oscillation
AMOC	Atlantic Meridional Overturning Circulation
AO	Arctic Ocean
AOSB	Arctic Ocean Science Board
ARW	Atlantic Return Water
ASOF	Arctic—Subarctic Ocean Fluxes
AW	Atlantic Water
AWI	Alfred Wegener Institute
BSO	Barents Sea Opening
CLIVAR	Climate Variability and Predictability
CTD	Conductivity, Temperature, Depth
DAMOCLES	Developing Arctic Modeling and Observing Capabilities for Long-Term Environmental Studies
D–O	Interstadial Dansgaard–Oeschger
EGC	East Greenland Current
EIC	East Icelandic Current
EOF	Empirical Orthogonal Functions
ENSO	El Nino Southern Oscillation
FS	Fram Strait
FSC	Faroe-Shetland Chanel
GPA	Geopotential Anomaly
GPS	Global Positioning System
GSA	Great Salinity Anomaly
GSDW	Greenland Sea Deep Water
GSG	Greenland Sea Gyre
GSP	Greenland Sea Project
GSR	Greenland–Scotland Ridge
ICES	International Council for the Exploration of the Sea
IOPAS	Institute of Oceanology Polish Academy of Sciences

IPCC	Intergovernmental Panel on Climate Change
JMC	Jan Mayen Current
KE	Kinetic Energy
LADCP	Lowered Acoustic Doppler Current Profiler
LSW	Labrador Sea Water
MIZEX	Marginal Ice Zone Experiment
MNAW	Modified North Atlantic Water
MOC	Meridional Overturning Circulation
NADW	North Atlantic Deep Water
NANSEN	North Atlantic-Norwegian Sea Exchanges
NAO	North Atlantic Oscillation
NCC	Norwegian Coastal Current
NML	No Motion Layer
NwAC	Norwegian-Atlantic Current
NwASC	Norwegian Atlantic Slope Current
OWSM	Ocean Weather Station Mike
PC	Principal Component
PIPS	Polar Ice Prediction System
PW	Polar Water
RAW	Return Atlantic Water
SAT	Surface Air Temperature
SB	Svalbard Branch
SD	Standard Deviation
SSH	Sea Surface Height
SST	Sea Surface Temperature
SVA	Specific Volume Anomaly
THC	Thermohaline Circulation
VEINS	Variability of Exchanges in the Nordic Seas
WSC	West Spitsbergen Current
WOCE	World Ocean Circulation Experiment

Chapter 1
Introduction

1.1 The Climate-Forming Role of the Ocean

The ocean has always influenced the Earth's climate system and the influence continues today. It is an extremely significant, if not essential, component of the system. For it is the ocean and life that originated there that created the atmosphere and has shaped its composition for billions of years. It is particularly important to speak about it now—in time of increased concern about the climate, when the out-of-control human impact upon the condition of the atmosphere is visible.

The ocean continues to be an active participant in the ongoing changes today as well, by transforming energy and matter significant in shaping of the global climate. It influences a very wide spectrum of processes, contributing to the atmosphere, lithosphere and cryosphere via a complex chain of relations. It is enough to mention that:

- Global hydrographic cycle is dominated by evaporation from the ocean which covers more than 71 % of the Earth's area and contains more than 97 % of water reserves on the Earth;
- The major part of sunlight reaching the Earth is absorbed in the surface layer of tropical oceans;
- The ocean gives up the heat by evaporation;
- The water vapour condenses and transfers latent heat to the atmosphere;
- In the tropical zone the ocean initiates the global large-scale atmospheric circulation by forming the Hadley cells of circulation;
- Condensation of water vapour at higher latitudes drives storms and winds;
- The ocean also heats up atmosphere by infrared radiation;
- The ocean controls the CO_2 contents in the atmosphere and dominates in the carbon cycle on the Earth;
- The oceanic phytoplankton has significant impact on cloud forming by the increase of SO_4 contents in the atmosphere;
- Sea ice modifies the water surface albedo;
- Denser deep water masses are formed during ice formation;
- The ice drift transports freshwater from polar regions towards the tropics;

W. Walczowski, *Atlantic Water in the Nordic Seas*,
GeoPlanet: Earth and Planetary Sciences, DOI: 10.1007/978-3-319-01279-7_1,

- The ocean stores and transports large amounts of heat through a system of sea currents.

The present work will focus mainly on the last item—the importance of the ocean circulation, and to be more precise—a part of it known as "thermohaline circulation", in redistribution of heat. Sea currents play an extraordinary role in the complex impact of the ocean on the climate. Thanks to very high heat capacity of water, its transport for long distances by sea currents is an unusually effective process of distributing energy excess from tropical regions towards the poles. The maximum northward heat transport on the North Atlantic ranges from 0.6 (Talley 2003) to 1 PW (10^{15} W) (Hall and Bryden 1982). At high latitudes, the incoming water may, having cooled down, sink to the deep basins and slowly return towards the equator, thus initiating the thermohaline circulation—a specific conveyor belt which carries heat and mass in the global scale.

1.2 Early Studies of the Ocean Circulation

When in 1751 the captain Henry Ellis, commander of the English slave-trading ship, found out that water taken from great depth is icy cold, he failed to appreciate the weight of that discovery. When sailing at latitude 25°N in the Atlantic Ocean, water was drawn from various depths with use of a custom-made 'bucket sea-gauge' and its temperature was read out. The captain wrote a letter to the originator of the experiment, a British clergyman—Stephen Hales, saying: "The cold increased regularly, in proportion to the depths, till it descended to 3,900 feet: from whence the mercury in the thermometer came up at 53 degrees (Fahrenheit); and tho' I afterwards sunk it to the depth of 5,346 feet, that is a mile and 66 feet, it came up no lower" (Rahmstorf 1999b).

That was the first ever recorded temperature measurement of the deep ocean, providing a result of great importance: deep ocean water is cold. Therefore, that was the first observation of the symptoms of a process which is today referred to as the Thermohaline Circulation (THC). In 1797, an American-born British inventor and physicist, Benjamin Thompson, published a correct explanation of that phenomenon: "It appears to be extremely difficult, if not quite impossible, to account for this degree of cold at the bottom of the sea in the torrid zone, on any supposition than that of cold currents from the poles; and the utility of these currents in tempering the excessive heats of these climates is too evident to require any illustration" (Rahmstorf 1999b).

The development of navigation contributed to better knowledge of the oceans. In 1832 a map of surface circulation of the Atlantic Ocean, drawn up by a surveyor in the Royal Navy and East India Company—James Rennell—was published (Richardson 2008). The map was the first to show the possibility of exchange between the North and South Atlantic as well as some components of the THC.

The development of theoretical studies followed the empirical recognition of the ocean. In 1845, Heinrich Lenz, a Russian physicist of Baltic German ethnicity, expanded the Thompson's idea by developing a conceptual model of two circulation cells, symmetrical in relation to the equator, including upwelling of water near the equator and downwelling in the Arctic and Antarctic regions, spreading of cold water at the bottom and compensatory warm poleward surface currents.

Great oceanographic expeditions in the second half of the 19th and the first half of the 20th century brought new maps and three-dimensional diagrams of the general ocean circulation. Such a diagram was developed after expeditions of the Challenger (1872–1876), Vivaldi (1898–1899) and Meteor (1925–1927). Concurrently, researchers became more and more aware of the role that the ocean circulation plays in shaping of the climate.

It was also then that the oceanographic research of the North Atlantic began, initiated by the countries of the northern Europe and Russia. In 1861, Carl Irminger, a Danish admiral, described a sea current which was then named after him. Danish expeditions discovered the Greenland-Iceland Ridge and Iceland-Scotland Ridge. In 1868, a British expedition determined Arctic and Atlantic water masses in the Faroe-Shetland Channel (FSC) for the first time.

Swedish and Russian expeditions explored waters around Spitsbergen. In the beginning of the 20th century, it was discovered that the West Spitsbergen Current (WSC) transports the Atlantic-origin water into the Arctic Ocean (AO) and that a portion of the water may recirculate southward together with the East Greenland Current (EGC). That discovery confirmed the origin of relatively warm water observed earlier in the East Greenland Current by Danish expeditions.

Between 1893 and 1896, Fram—a custom-design ship, constructed to withstand the winter sailing conditions and capable of wintering in ice—went on an expedition organized by Fridtjof Nansen. The expedition provided essential information about the Arctic Ocean, geography and hydrography. During the cruise, Nansen got convinced about extraordinary weight of exchange between the North Atlantic and the Arctic Ocean.

In 1900 Norwegian researchers received the research vessel "Michael Sars". Based on the 4-year work aboard that ship, Helland-Hansen and Nansen (1909) presented a complex description of the oceanographic conditions of the Nordic Seas (NS). The quality and the accuracy of the results which became a major source of information about physical oceanography of those regions amazes even today.

In 1905, Nikolai Knipovich, a Russian researcher, described very cold and salty bottom water of near-freezing temperature as a product of salt rejection during sea ice formation at the Barents Sea.

Based on the observation of cold bottom water in the Greenland Sea, in 1909, Nansen was the first one to present the classical model of the Greenland Sea Deep Water (GSDW) formation. That is how he initiated a debate over the origin and importance of GSDW and the process of deep convection, which has been continued for nearly 100 years until now.

In the early 1920s, the idea of climate change caused by the changing ocean circulation was formulated by Thomas Chamberlin, an American geologist.

After an active period in the beginning of the 20th century, the research of the Nordic Seas suffered a period of stagnation. Only in the period 1935–1936 some data were collected in the Norwegian Basin, the next series of research were not performed until the 50s. That winter data did not bring the expected results—no homogenous and cold water column was found, therefore, it was assumed (Metcalf 1955; after Blindheim and Østerhus 2005) that the deep convection was not a very important phenomenon.

1.3 Formulating the Idea of Thermohaline Circulation

In the 50s, Henry Stommel, an American oceanographer conducted pioneer works on three-dimensional structure of the ocean and the ocean circulation. In 1957 he revolutionized the concept of the Atlantic circulation, and in 1961 he was the first one to introduce the concept of the thermohaline circulation (Stommel 1961).

Oceanographic research intensified during the International Polar Year 1958. Among other things, that complex research project included widespread studies of the Arctic—Polar Front Survey. Being aware of the significance of the dense deep water, a lot of effort was put into clarifying the origin of the GSDW. Since it was impossible to observe a homogenous water column in the Greenland Sea in winter, Carmack and Aagaard (1973) suggested that GSDW may form at medium depth during the double diffusion process.

Since 1970 a lot of research programmes have focused on the Greenland Sea. In February 1971 a nearly homogenous water column was observed in the central part of the Greenland Sea. On the initiative of the International Council for the Exploration of the Sea (ICES), complex data for the research of GSDW formation were gathered in winter and spring 1982 (Clarke et al. 1990). The Greenland Sea Project (GSP), implemented between 1987 and 1993, also contributed to recognition of the deep water formed in the Arctic Ocean as a component of deep water of the Nordic Seas.

During that period, the research in the Fram Strait (FS) focused on the processes occurring at the ice margin. The largest project was the international Marginal Ice Zone Experiment (MIZEX) in 1984. It provided a lot of interesting research material, indicating, inter alia, the role of mesoscale vortices in heat transport and ocean-ice interaction.

The most interesting phenomenon at the southern boundary of the Nordic Seas, drawing attention of oceanographers was the overflow of the Arctic water above the Greenland-Scotland Ridge. The process was described for the first time in 1898 by Knudsen, based on the results of the Danish “Ingolf” expedition dated 1895–1896. In 1960 ICES coordinated the international research of the Greenland-Scotland Ridge, known as the Overflow Experiment. The main result of the programme was the conclusion that the overflow occurs in nearly every place where the depth above the ridge exceeds 300 m.

In 1985 ICES initiated the NANSEN project (North Atlantic-Norwegian Sea Exchanges) to investigate the exchanges of volume, heat and salt above the Greenland-Scotland Ridge. Also in the northern part of the WOCE programme (World Ocean Circulation Experiment)—Nordic WOCE, the exchange between the North Atlantic and the Nordic Seas was investigated. The works were continued under the European Union's VEINS (Variability of Exchanges in the Nordic Seas) program.

However, it is only 100 years after Thompson's work, that, in 1987, the Conveyor Belt Circulation theory was formulated. Wallace Broecker (1987), an American researcher, presented a concept of surface and deep sea currents combining all the oceans and transporting salt and heat among them (Fig. 1.1). That revolutionary idea was equally supported and opposed. Similarly to Chemberlin, Broecker also claimed that changes of the Atlantic thermohaline circulation had led to violent and widespread changes of the climate, which had occurred in the North Atlantic region during recent glaciations.

Further years brought numerous clones of the complex circulation idea, covering the world ocean area, revised and improved diagrams and new terminology. The debate on what drives the circulation and what processes play the most important role has been continued up to the present day. Finally, in the period of visible climate warming, a widespread discussion has begun on the role of the ocean currents in climate forming, particular the significance of the conveyor belt circulation, and even the risk of its modification, slow-down or even shutdown. The Arctic Ocean, including the adjacent seas, which is a key region for the circulation, has become the crucial object of study for oceanographers, meteorologists and climatologists.

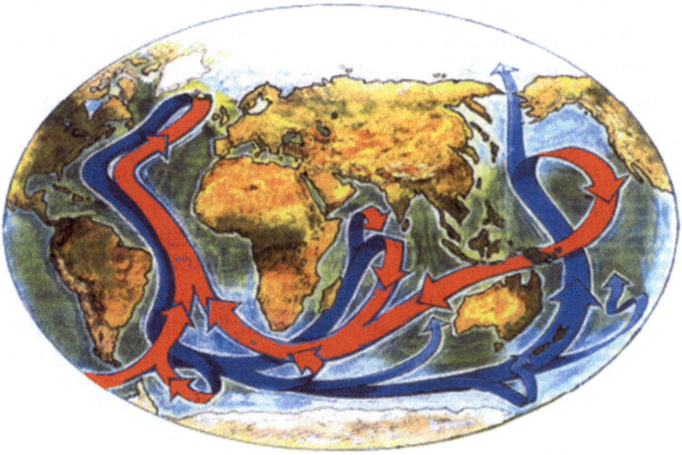

Fig. 1.1 Broecker's (1987) scheme of the global thermohaline circulation

1.4 Modern Investigations of the Thermohaline Circulation

In 1999, the Arctic Ocean Science Board (AOSB) announced plans to study the two-way exchanges and links of the Arctic Ocean and the subarctic seas. The reason for that was the fact that a lot of climatologists indicated the possibility of weakening of the thermohaline circulation due to the release of increasing quantity of greenhouse gases into the atmosphere. The North Atlantic was considered as the most sensitive area, where the inflow of freshwater from the melting Arctic ice could enhance the stratification of surface water in the Greenland Sea, disturb the deep convection process, weaken the flow of deep water over the Greenland-Scotland Ridge and the North Atlantic Deep Water (NADW) formation. The data collected previously during the VEINS project were not sufficient enough either to directly support or negate that thesis. The values of heat and volume transports were determined, nevertheless, the time series of the measurements were too short to define their variability. Furthermore, the measurements failed to include directly the Greenland-Scotland Ridge and the overflow zone. The European Union's Arctic-Subarctic Ocean Fluxes (ASOF) program, which commenced in 2002, was supposed to fill that gap. The direct objective of the ASOF program, similarly to the VEINS, was to investigate the variability of flows between the Arctic Ocean and the Atlantic Ocean by measurements and modelling. The long-term objective was to develop a system of measurements enabling determination of the role of the Arctic Ocean and the subarctic seas in decadal climate changes. The ASOF program is currently continued as DAMOCLES (Developing Arctic Modelling and Observing Capabilities for Long-term Environmental Studies)—the integrated project of studying and monitoring of the atmosphere–ocean-ice system, which purpose is to observe, comprehend and evaluate the climate change in the Arctic.

The research of climate variability, interactions between the ocean and atmosphere, are conducted under international cooperation. One of the most complex programs is CLIVAR (Climate Variability and Predictability). The program was initiated in 1998 to study variability and predictability of climate on time scales ranging from months to decades, as well as the reaction of the climate system to anthropogenic factors.

The Institute of Oceanology of the Polish Academy of Sciences (IOPAS) is an active participant in the numerous above-mentioned science projects. Since the beginning of our work in the Arctic, the research has been conducted in cooperation with other institutions. We have been taking part in international programs, concurrently maintaining our own profile of research, adjusted to the specific characteristics of the Institute's ship and the constraints of the measuring period. The research programs conducted at IOPAS are described in more detail in Chap. 3.

1.5 Contemporary Views on the Thermohaline Circulation

The short historical background presented above shows the magnitude of evolution in oceanography and climatology in recent years. Despite the fact that the first ideas about the Arctic origin of deep water and the significance of the thermohaline circulation for the Earth's climate had been formulated more than 200 years ago, it is only recently that we appreciated the weight of those processes. As researchers, we participate in the process of recognizing the ocean as an important climate-forming factor. Simultaneously, we observe the growing social awareness, the interest in the climate issues. It is a side effect of the global warming—the phenomenon which has been one of the main subjects of discussion on all forums since the Nobel prize was awarded for achievements in disseminating knowledge about it.

Current views of the Conveyor Belt are different than those presented by Broecker. At present, when describing the global ocean circulation, the term thermohaline circulation is used, and in the case of south-north transport in the Atlantic Ocean—the Meridional Overturning Circulation (MOC) or the Atlantic Meridional Overturning Circulation (AMOC). The role of the Atlantic Ocean, particularly the processes in its northern regions—the Nordic Seas and the Arctic Ocean—is specifically exposed as essential in maintaining the THC (Hansen et al. 2004).

According to the current state-of-the-art and views on climate phenomena, the THC is the main driving mechanism for sea currents (Rahmstorf 2003). Water cooling and ice formation at high latitudes increases the density of surface water sufficiently enough to initiate the deep convection which, along with other processes, is referred to as the thermohaline ventilation (Fig. 1.2). On the northern hemisphere, deep water is formed mainly in the northern areas of the Atlantic Ocean, separated from its southern part by the Greenland-Scotland Ridge (GSR). 75 % of the Atlantic Water flowing over the ridge into the Arctic is transformed into deep dense water (Hansen and Østerhus 2000). On the opposite end of the circulation loop, in the tropical zone, vertical mixing processes cause an increase of the deep water temperature, reducing density and cause its upwelling to the surface. As a result of cooling and ventilation at high latitudes, as well as heating and vertical mixing at low latitudes, a horizontal differences in water density occurs. These differences force surface flow of warm and salty water towards the pole and deep flow towards the equator. The main water mass formed in the chain of ventilation and flow processes over the Greenland-Scotland Ridge is the North Atlantic Deep Water which feeds most of the deep water of the global ocean (Hansen et al. 2004).

However, the above simplified description of the THC functioning has its opponents (Wunsch 2002; Wunsch and Ferrari 2004). Discrepancies of the views result from the fact that the THC is a lot more complex than as described in the above scheme. All its phases are linked by a chain of feedbacks, which makes it difficult to perform a precise cause and effect study. The thermohaline ventilation

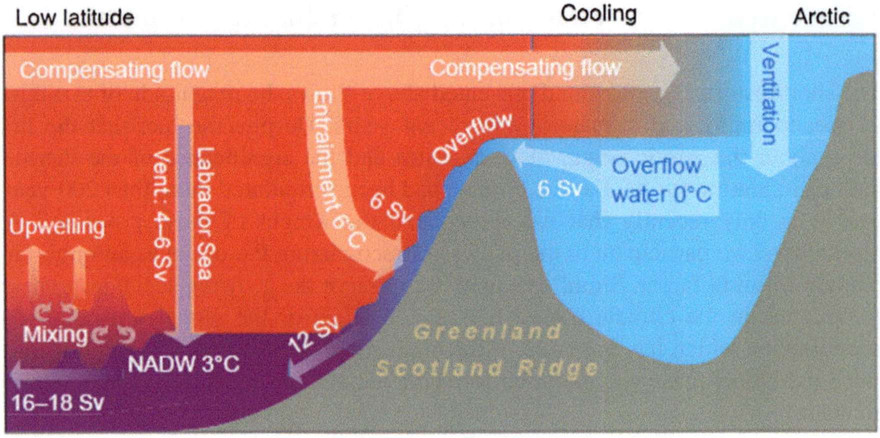

Fig. 1.2 AMOC. The formation of NADW (Hansen et al. 2004)

processes are considered as the most essential. It is the convection in the Arctic that generates horizontal differences in density and pressure, which drive the exchange between the Arctic and Atlantic Oceans. Due to high salinity, surface waters in the North Atlantic, in spite of mixing with the surrounding, less salty water masses, having cooled down, gain sufficient density to sink into the depths, thus replacing waters which reside there, and—to picture it more clearly—they make room on the surface for waters from the south to arrive. Hence the name of the process—thermohaline circulation (from Greek *thermos*—heat, and *haline*—salt), since those two primary physical properties of sea water: temperature and salinity, play equally important role in the circulation. The phenomenon of stable circulation of the World Ocean is frequently compared to the operation of a thermodynamic machine driven by the heating/evaporation cycle at low latitudes and cooling/condensation at high latitudes (Rahmstorf 2006). There is a certain analogy to the atmospheric circulation—the Hadley cells—in that view.

The most important difference between the atmosphere and the ocean is the fact that, unlike the atmosphere, the ocean is heated and cooled only on the surface, and stable vertical stratification occurs nearly over its entire area. The thermohaline ventilation—the places of contact between the abyssal and the surface may occur only in specific, restricted regions. On the Northern Hemisphere it is the Labrador Sea and the Greenland Sea, and on the Southern Hemisphere, the water ventilation occurs near the Antarctic-in the Ross Sea and the Weddell Sea (Fig. 1.3). Despite the link with the Arctic Ocean and sufficient cooling, there are no places where deep convection occurs in the Pacific Ocean, since the salinity of the surface Pacific Water is too low (32) to reach the deep water density. The Indian Ocean is closed at the north, therefore, it cannot participate in the described processes.

The places where surface water sinks into the deep basins are a sort of flywheels which drive the entire THC. The potential energy of water masses is transformed

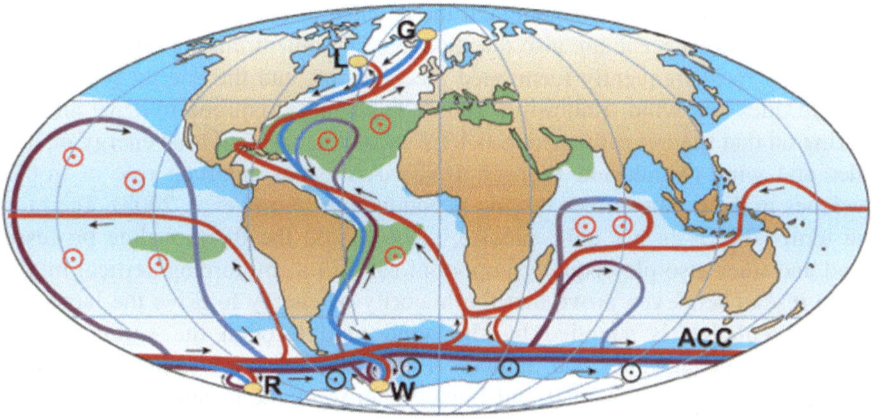

Fig. 1.3 Scheme of the global thermohaline circulation. Surface currents—*red line*, deep currents—*blue line*. The main areas of deep-sea water formation in the Greenland (*G*), Labrador (*L*), Weddell (*W*) and Ross (*R*) Seas are outlined in *yellow*. Areas of mixing and upwelling are ringed (Kuhlbrodt et al. 2007)

here into the kinetic energy of flows. As regards forming of the THC and the climate on our hemisphere, the North Atlantic and the Nordic Seas are the most important regions, and the Greenland-Scotland Ridge, is the most important bottom area. The flows occurring over the ridge—southward deep water flows and the compensating northward surface flows—play critical roles in the magnitude of volume and heat exchange between the Atlantic Ocean on one side and the Nordic Seas and the Arctic Ocean on the other.

In addition to the vertical water exchange (convection) and the currents it drives, there is also another type of circulation which is important in the Arctic—horizontal estuarine type of circulation. The incoming warm and saline Atlantic Water mixes with fresh surface water in the Nordic Seas and the Arctic Ocean, and then returns to the south, cooled down and less saline (recirculation). Besides that, there is also the cold and low-saline East Greenland Current, flowing southward from the Arctic Ocean. The two types of circulation, creating unique ocean currents system and conditions in the Arctic, are sensitive to climate change (Loeng et al. 2005).

The concept of thermohaline forcing of the oceanic circulation is widely accepted, although there are some differences in views as to what elements are essential. However, there is a lot of negative opinions on its complexity or even validity. The most popular objection concerns the non-conformity of the simple THC pattern with Sandström theorem which claims that cooling and heating of a sea region on its surface will not lead to the occurrence of steady circulation; the source of heating must be below the cooled layer. Although the areas and mechanisms of cooling and sinking of surface waters are known, the tropical end of the THC is much less recognized. The patterns and regions of warm water mixing are not known, the source and the balance of energy required to transport heat into the deep ocean are also questionable. Turbulent mixing (Huang 1999) is one of the most frequently mentioned mechanisms, and the energy sources may include tides and wind (Rahmstorf 2006). The energy required to drive turbulent mixing on that scale is estimated as 0.4 TW, which is 20 % of total energy input by tides and wind (Munk and Wunsch 1998). Also, the important role of tropical cyclones in the ocean surface mixing is pointed out (Korty et al. 2008). Therefore, the term—THC—is not quite precise, since beside the thermohaline processes, wind and tides also play equally important roles in it, by forcing vertical mixing. Some researchers even prove that the majority of energy to drive the large-scale ocean circulation is supplied by the wind field through the vertical mixing mechanisms (Toggweiler and Samuels 1995). The energy of winds and tides is transformed into the potential and kinetic energy of the ocean, thus driving the large-scale circulation, and the thermohaline effect might be even slowing it down.

A number of misunderstandings and inaccuracies have arisen as regards the THC. It is often the case that wind-driven currents, such as the Gulf Stream, are mistaken for the THC. Both of the circulations overlap each other in some areas and export the Atlantic Water northward. However, their range, transport routes and, above all, driving mechanisms are different. The THC is often identified with the AMOC and both terms are used interchangeably. The THC is mostly an idea based on physics, including the complex ocean circulation driven by thermohaline forces. However, AMOC is a total northward/southward flow in which driving forces are not distinguished. Unlike the THC, the upper part of AMOC includes wind-driven currents in the Ekman layer and the deep layer is driven by overturning. Thus, the idea of AMOC includes, in a simplified way, both the wind-drive circulation and the thermohaline circulation. The northern part of AMOC is estimated to generate a flow of 17 Sv (1 Sverdrup, Sv = 10^6 m^3/s) (Talley et al. 2003), whereas the southern part—14 Sv (Orsi et al. 2002).

Some authors claim that the THC is just a useful theory helping to distinguish—especially in numerical modelling—between various types of flows. However, it is difficult to accept that interpretation. The THC has been proven to be a significant physical phenomenon. This is confirmed by the fact that the mean air temperature in the areas of deep water formation is up to 10 °C higher than the mean air temperatures for given latitudes. This refers both to the northern and the southern hemisphere (Rahmstorf 2000).

1.6 The Arctic in the Global Climate System

The Arctic, including in particular the Nordic Seas, is linked to the global climate system mainly by balancing the heat received by the system at low latitudes and given up at high latitudes. The heat is accumulated and transported by the atmosphere and the ocean. Only 3–5 % of the heat supplied by the ocean–atmosphere system to the polar regions comes from the ocean advection (Bobylev et al.

2003), however, the ocean processes are extraordinarily significant in shaping the climate of the Arctic, which cannot be reduced to the above mentioned percentage value.

The ocean's heat capacity is 1,000 times greater than that of atmosphere. The heat accumulated at low latitudes is stored, transported and given off in another region, often with a delay of many years. Most of the heat is accumulated in the upper 700 metres of the ocean, however, a portion of it is transported from the surface water to deep layers. At the latitude of 30°N, currents of all the oceans carry ca. 2 PW of sensible heat poleward from the tropics (Stewart 2006). The value is equivalent to the atmospheric transport. At the latitude of 60°N latitude 0.4 PW is transported towards the Arctic, mainly by the warm and highly saline Atlantic Water. During the northward transport of the AW, heat fluxes of ca. 200 W/m^2, given off by the ocean in winter, heat up the atmosphere and modify its circulation. Due to that, the AW carries along much less heat to the Arctic Ocean than to the Nordic Seas. Schauer et al. (2008) estimate the value of the oceanic heat transport toward the AO through Fram Strait as 16–41 TW (1 TW $= 10^{12}$ W). However, even that reduced quantity plays a decisive role for the equilibrium of the climate system.

The atmosphere transports heat mainly in water vapour as latent heat. At high latitudes it is released as sensible heat during the water vapour condensation. During the process, precipitation and river discharge additionally supply freshwater. In the Arctic the water accumulates in the surface layer and in the halocline—a salinity gradient layer—and is afterwards exported southwards. Through mixing with sea water and transport towards tropics, the freshwater is reintroduced into the global hydrographic cycle. This prevents the increase of salinity in the tropical ocean due to evaporation. The freshwater is exported from the Arctic mostly by cold, low-saline sea currents, such as the East Greenland Current, and the flow through the Canadian Arctic Archipelago, but also by drifting sea ice. Intermediate waters that form at the boundaries of fronts and in vortices, as well as deep waters that form on the Arctic Ocean shelves, being less saline than the Atlantic Water, also export freshwater southwards and participate in the hydrographic cycle.

The unique role of the Arctic is thus connected with the thermohaline circulation for which it is essential that heat and freshwater is transported by both the atmosphere and the ocean. The THC is very much dependent on the export of freshwater from the Arctic. The supply of freshwater from the Arctic Ocean, which is intensified (or has been intensified) as a result of the warming, will probably increase stratification of the north Atlantic Ocean and the Nordic Seas, which may in turn hinder NADW formation and weaken the THC. This will result in much lower northward inflow of the Atlantic Water, slowing of AMOC and thereby cooling down. The leading hypothesis explaining the abrupt climate changes which have taken place over the past millennia points at the changing quantities of the formed NADW and, consequently, the strength of the THC. The research which results were published in 2006 (EPICA 2006) seems to confirm that. Thanks to a new ice core dating method, scientists successfully synchronized in time the cores sampled in the Arctic (Greenland) and the Antarctic. The obtained data showed that each temperature rise on Greenland was correlated with temperature drop in the Antarctic. The contemporary climate models based on the AMOC idea provide such a relationship. The increase of temperature on Greenland is connected with intensification of the AMOC, greater transport of heat from the tropics towards the North Atlantic and cooling down in the south.

Sea ice is an inseparable component of the Arctic. The ice cover isolates water from the atmosphere, regulates the exchange of energy and moments in the system. Brine is rejected during ice formation and the sea ice transports freshwater by drifting. The range, surface, thickness and seasonal variability of sea ice are extremely important climate factors.

The above short description demonstrates how important is the Arctic in the global climate system, particularly the Arctic Ocean and the Nordic Seas. On one hand they are climate-forming regions, whereas on the other, they are very much vulnerable to any changes and deviations of the climate. Polar regions are thus a great indicator of the ongoing climate change.

1.7 Climate Changes in the Arctic

There are two types of climate change. The first one results from the change in the heat budget of the entire planet, which is caused, for instance, by a variation of the incoming quantity of the solar radiation, a change of the albedo, or a change of the long-term thermal radiation (greenhouse effect). In that case a global change occurs. The second type is a change of climate caused by a change of heat redistribution by the atmosphere and the ocean. The same quantity of the solar heat reaching the Earth is distributed in a different way, thus producing abrupt changes which affect large areas. The changes occur in a bipolar system: warming on the Northern Hemisphere is correlated with cooling on the Southern Hemisphere.

The Intergovernmental Panel on the Climate Change announced that within a time of 100 years, from 1906 to 2005, the mean Surface Air Temperature (SAT) over the land rose by approximately 0.74 °C (IPCC AR4 2007). The two recent periods of noticeable climate warming are the 1910–1949 period and the period which began in the 1970s and lasts until the present time (Fig. 1.4).

A widespread debate has been going on with reference to those changes. The latest IPCC reports leave no doubts as to the anthropogenic background of the changes, observed greenhouse effect caused by the increased emission of greenhouse gases in the industrial age. Therefore, that would be the first type change, mainly caused by the global change in the heat balance. The ocean certainly plays a significant, though not yet clearly recognized, role. The great thermal inertia of the ocean can even weaken the effects of the climate change and delay them.

However, some voices can be heard against the anthropogenic explanation of the observed warming. The opponents remind that there have been great coolings and sudden warmings in the history of the Earth. The most frequently mentioned example is Greenland, since the climate of the glacial periods of the North Atlantic

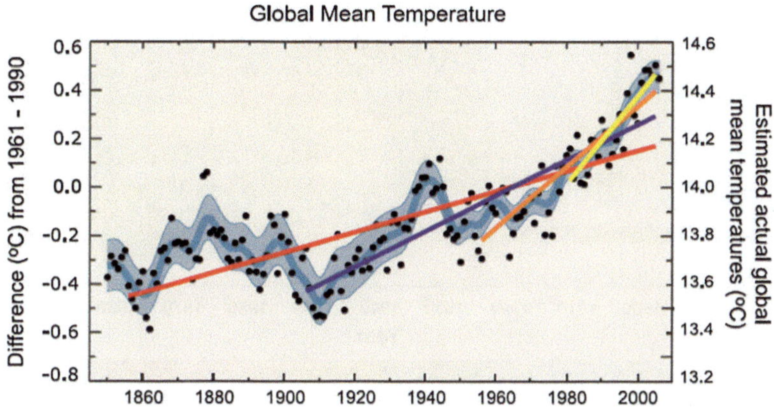

Fig. 1.4 Mean annual air temperatures (*dots*), smoothed data (*blue line*) and linear trends for different periods of time (*coloured lines*). Left-hand axis—temperature anomalies in relation to the mean for 1961–1990, right-hand axis—real temperatures (after IPCC AR4 2007)

has had some very long cool periods and shorter and intense warmings. During the Dansgaard-Oeschger (D-O) interstadial, the temperature on Greenland rose even by 8–16 °C in a few decades and then smoothly returned to its mean level. Thus, the current warming is often compared to the D-O event.

After the period of glaciations, approximately 24,000 years ago, the process of transition to the interglacial began. Changes of temperatures were less rapid than currently, the warming in Greenland advanced at an average rate of 2 °C/ 1,000 years. Climate similar to the present one formed approximately 10,000 years ago, and the climate optimum occurred 9,000–5,000 years ago. Most of glaciers covering the north Norway and Spitsbergen melted then, and the tree line shifted far north in Norway. Mean annual temperature was circa 2 °C higher than the average for the period 1961–1990. The main driver of that climate optimum was very high solar radiation in summer. At 60°N latitude, it exceeded the present insolation by as much as 8–10 % (Nesje et al. 2005).

During the two recent millennia, the climate of the Nordic Seas was also subjected to some oscillations, including both slow warming and abrupt coolings. The reconstructed Sea Surface Temperature (SST) shows relatively constant values until the year 700, and then, in the period 700–1,000, some episodes of gradual temperature increase by 0.5–1 °C per 100 years. The period 1,200–1,400 is called the Medieval Warm Period, however, it was interrupted by sudden coolings. After that, a cooler period occurred, including the Little Ice Age, between 1,550 and 1,920, when the annual mean temperature was 1 °C lower than the 1901–1995 long-term mean temperature (Luterbacher et al. 2004). The next significant warming at high latitudes happened between 1920 and 1940. The cooling which followed it lasted until the early 1970s, when the current period of temperature increase began (Overland et al. 2004; Serezze and Francis 2006).

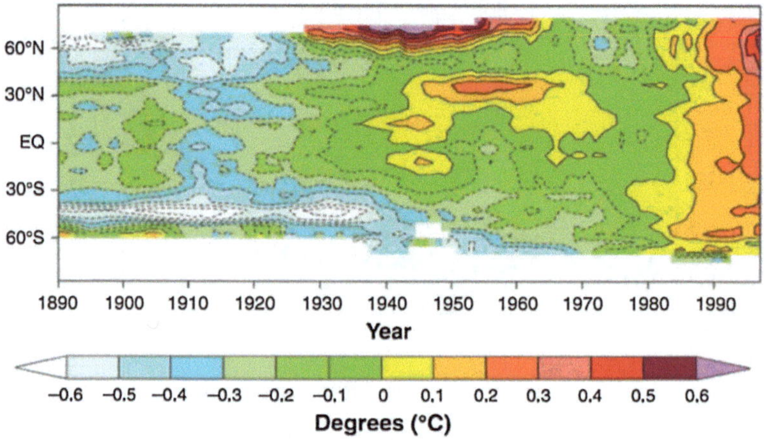

Fig. 1.5 Observed, zonally averaged SAT [°C] anomalies as a function of latitude and time (Alley et al. 2003)

Both of the recent warmings are frequently compared and the example of the 30s is an argument to support the thesis that the present warming is a natural process, since in the 30s the concentration of the anthropogenic greenhouse gases in the atmosphere was much lower than currently. However, the primary difference betwen the both warmings lies in their range. While the first one covered mainly the Arctic and featured changes in redistribution of heat with the same heat balance of the entire planet, the warming which is currently observed concerns the entire globe (Fig. 1.5). It can be observed both on the northern and southern hemisphere. As a result of changing conditions of long-wave radiation caused by accumulation of greenhouse gases in the Earth's atmosphere, more and more heat is stored in the global climate system. The post-industrial increase in the quantity of greenhouse gases caused average rise of the absorbed radiation of ca. 2.45 W/m^2. This is merely 1 % of the incoming solar radiation (Hardy 2003), nevertheless the global temperature rise may reach as much as 3–5°C in a century. Such an increase has been never recorded in the previous history of the Earth.

1.8 Causes and Effects of Changes in the Thermohaline Circulation

Regardless of the origin of current warming, the thermohaline circulation is a significant heat-redistributing and climate-forming element. Its intensity changes as the NADW production changes. The key question which still needs to be answered is why the production changed in the past and how it will change in future.

There are some opinions claiming that the variability of deep water formation is a natural, internal oscillation of the ocean–atmosphere climate system. Such oscillations are known in nature—the most well-known is the El Niño Southern Oscillation (ENSO) in the tropical Pacific, the Arctic Oscillation, the North Atlantic Oscillation (NAO) or the Atlantic Multidecadal Oscillation (AMO). Proshutinsky and Johnson (1997, 2001) described oscillations in the Arctic Ocean circulation. Dukhovskoy et al. (2004, 2006a, b) presented the exchange between the Arctic Ocean and the Nordic Seas as a process controlled by the system's natural oscillations.

However, the oscillations are short-term, quasi-decadal changes in the regimes of atmosphere and ocean circulation and they do not provide explanation for long-term climate changes. Obviously, longer cycles related to solar activity or changes of the Earth's orbit are also known to occur in nature. One of the best known is the Milankovitch cycle which explains the changes in the Earth's insolation by the superposition of three natural oscillations of the Earth's orbit: the tilt of the Earth's axis related to the ecliptic plane, precession of the rotation axis and eccentricity of the Earth's orbit.

Model research shows (Ganopolski and Rahmstorf 2001) that THC is an extremely sensible phenomenon and it reacts in a non-linear way—a change from the steady "cool" type of circulation into the steady "warm" are sudden. The research does not directly reflect the present situation, however, they show that there is a critical value of freshwater in the system which caused the change in the THC regime in the past. Too low salinity of the surface water in the deep water formation areas may slow down or even bring the process, which is the main driver of the THC, to a halt. There is evidence from paleoclimate research and numerical models (Rahmstorf and Ganopolski 1999) that glaciations were caused by a disturbance in the North Atlantic freshwater budget. Furthermore, lesser climate oscillations, periodical coolings or warmings were caused by the change of heat flux carried northward by the THC surface currents (Clark et al. 2001).

In the age of instrumental observations of the North Atlantic, the Nordic Seas and the Labrador Sea, numerous changes of salinity have been observed caused by the change in the freshwater inflow and changes of atmospheric circulation—mainly the NAO index. The largest anomaly, called the Great Salinity Anomaly (GSA), was observed in the period 1968–1978, whereas others occurred in the 1980s and 1990s. The anomalies affected the AMOC by production of the Labrador Sea Water (LSW), the main component of the NADW. During the positive NAO phase, the LSW production increases (Dickson et al. 1996).

Variations of the THC intensity are also related to the Atlantic Multidecadal Oscillation (Knight 2005). The AMO is characterized by variations of the surface temperature of the North Atlantic in the range of 0.4 °C with a period of 65–75 years. A warm phase began in the early 1990s, and in 2005 the tropic Atlantic temperature reached record high values. According to Knight, the THC is currently in the phase of natural maximum, and its intensity will begin to decrease in 2020. Another American oceanographer, Harry Bryden, thinks that we are currently observing major changes of the THC. In the article summarizing the

RAPID program, which monitors and models the AMOC on the basis of measurements of the MOC taken at 25°N, the authors prove that the AMOC slowed down by about 30 % between 1957 and 2004 Bryden et al., 2005. Whereas the surface northward Gulf Stream transport remained unchanged, the southward transport of the NADW decreased by 50 %. In 2004, most of the Gulf Stream water was recirculating back southward within the subtropical gyre. Bryden's study was a sensation among oceanographers and climatologists. Even the originators of the idea stating that the AMOC is slowing down or coming to a halt claim that the decreased transport as observed by Bryden is an artefact and not a real trend, and climate change has not yet reached magnitude necessary to trigger such disastrous processes (Schiermeier 2006). On the other hand, the analysis of Cunningham et al. (2007) shows that natural AMOC variability is strong enough to yield results similar to those observed by Bryden. The debate served as an opportunity to refer back to the primary issue—the necessity of obtaining sufficiently long and accurate time series. The estimate of transport based on sporadic measurements of sea currents which served as a basis for the Bryden's study includes a major error, since the short-term variability of the transport is much greater than assumed until now.

Results of modelling conducted under the RAPID program predict that the AMOC will slow down in the 21st century and then recover after 2100, provided that CO_2 concentration in atmosphere is held fixed (Steynor and Wallace 2007). However, neither the RAPID models nor other climate models predict that the AMOC will stop. The potential results of cooling caused by the AMOC slowing down will be compensated by warming caused by the greenhouse effect.

1.9 The Objective and Scope of this Work

We are living in times of quick climate change, evolving in front of our eyes. We feel the change, we study it, however, we cannot have any significant impact on it. We also participate in another process—evolution in recognizing and understanding the role of the ocean in the global climate system. In that case we are not only passive observers of that process, we take active part and form it. By commencing measurements in the Arctic in 1987, oceanographers from the Institute of Oceanology of the Polish Academy of Sciences in Sopot entered the stream of modern oceanographic research, closely related to the study of climate and its transformation. For years our comprehension of the ocean have evolved.

The primary objective of the work is to integrate, analyse and present the results of the research conducted in the Nordic Seas by the Physical Oceanography Department in the context of the impact which the studied processes of the changes in the Atlantic Water properties and heat transport by the West Spitsbergen Current have upon climate. The period between 2000 and 2007 was considered to represent an incident in the latest history of that region sudden warming and then cooling of the Atlantic Water, and the emphasis was put on the data from that period. The work focuses on the Atlantic Water, its variability, large-scale processes of heat transport by the AW volume.

The section presenting the IOPAS data follows a short description of problems connected with the Earth's climate, the role of the ocean in shaping the climate and changes in the Arctic climate. Further, the description of bottom topography, circulation and water masses in the Nordic Seas is provided. In the subsequent section of the work, the methodology of measurements is presented along with the analysis of the collected data, while concentrating on the characteristics of the Atlantic Water and estimation of the heat transported with it by the West Spitsbergen Current.

In order to be able to predict future changes of the AW properties, the Arctic climate and, consequently, the world climate, the structure and variability of the WSC must be recognized, as well as the variability of the Atlantic Water properties. The work includes a study of its physical and biogeochemical properties, spatial distribution, variability in summers 2000–2007. The velocities of baroclinic currents and baroclinic transports of masses and heat were calculated. The structure of the West Spitsbergen Current is described as well as the temporal variability of the baroclinic flow field. The Atlantic Water transformation processes are described as well as changes of its physical and chemical properties during the northward advection. The author sought relationships which would allow to predict the AW properties in the Fram Strait area, based on the knowledge of the AW properties in the southern part as well as its approximate advection velocity.

A series of calculations was conducted during the work, and having obtained new measurement data, the previously formulated statements were verified. Since a lot of results were obtained and ideas examined, they are discussed on an ongoing basis, and respective conclusions are formulated with a lot of caution, particularly considering the fact that they sometimes contradict each other. That type of approach may differ from a widely adopted form of scientific theses, however, one must bear in mind that the research is performed upon complicated, inter-connected processes on the basis of short time series, and it is not always possible to obtain unequivocal answers. Thus, Ockham's razor principle was not applied in each case; it was used only when drawing final conclusions. It is also worth remembering that references concerning the studied processes often contain contradicting descriptions, results, conclusions and opinions. In addition to the difficulties mentioned above, the results gained from numerical models differ significantly. That reflects our limitations in understanding how the large-scale climate system operates (Stouffer et al. 2006).

Recent observations show warming of the North Atlantic waters (Holliday et al. 2007, 2008), intensification of the westerly winds, reduction of deep water formation in the Greenland Sea (Ronski and Budeus 2005), dramatic reduction of the Arctic Ocean ice coverage (Comiso et al. 2008; Rothrock et al. 2008; Parkinson and Cavalieri 2008). Thus, a question arises at this point—are we witnesses of a global change, or is it a manifestation of long-term cycle occurring in the complex ocean–atmosphere system?

References

Alley RB, Marotzke J, Nordhaus WD, Overpeck JT, Peteet DM, Pielke RA Jr, Pierrehumbert RT, Rhines PB, Stocker TF, Talley LD, Wallace JM (2003) Abrupt climate change. Science 299:2005–2010

Blindheim JV, Østerhus S (2005) The nordic seas, main oceanographic features. The Nordic Seas. An Integrated Perspective 158:11–39 AGU Geophysical Monograph

Bobylev LP, Kondratyev KY, Johannessen OM (2003) Arctic environment variability in the context of global change. Springer-Verlag, Berlin, p 471

Broecker WS (1987) The biggest chill. Natural History, pp 74–82

Bryden HL, Longworth HR, Cunningham SA (2005) Slowing of the Atlantic meridional overturning circulation at 25°N, Nature, 438:655–657, doi:10.1038/nature04385

Carmack E, Aagaard K (1973) On the deep water on the Greenland Sea. Deep-Sea Research 20:687–715

Clark P, Marshall JS, Clarke GK, Hostetler SW, Licciardi J, Teller J (2001) Freshwater forcing of abrupt climate change during the last glaciation. Science 293:283–287

Clarke RA, Swift JH, Reid JL, Koltermann K (1990) The formation of Greenland Sea Deep Water: double diffusion or deep convection? Deep-Sea Research 37(9):1385–1424

Comiso JC, Parkinson CL, Gersten R, Stock L (2008) Accelerated decline in the Arctic sea ice cover. Geophys Res Lett 35:L01703. doi:10.1029/2007GL031972

Cunningham S, Kanzow T, Rayner D, Baringer MO, Johns WE, Marotzke J, Longworth HR, Grant E, Hirschi JJ-M, Beal LM, Meinen CS, Bryden HL (2007) Temporal variability of the atlantic meridional overturning circulation at 26.5°N. Science 317(5840):935–938, DOI: 10.1126/science.1141304

Dickson R, Lazier J, Meincke J, Rhines P, Swift J (1996) Long-term coordinated changes in the convective activity of the North Atlantic. Prog Oceanogr 38:241–295

Dukhovskoy DS, Johnson MA, Proshutinsky A (2004) Arctic decadal variability: an auto-oscillatory system of heat and freshwater exchange. Geophys Res Lett 31:L03302. doi:10.1029/2003GL019023

Dukhovskoy D, Johnson M, Proshutinsky A (2006a) Arctic decadal variability from an idealized atmosphere-ice-ocean model: 1. Model description, calibration, and validation. J Geophys Res 111:C06028. doi:10.1029/2004JC002821

Dukhovskoy D, Johnson M, Proshutinsky A (2006b) 'Arctic decadal variability from an idealized atmosphere-ice-ocean model: 2. Simulation of decadal oscillations. J Geophys Res 111:C06029. doi:10.1029/2004JC002820

EPICA-COMUNITY-MEMBERS (2006) One-to-one coupling of glacial climate variability in Greenland and Antarctica. Nature 444(9), doi:10.1038/nature05301

Ganopolski A, Rahmstorf S (2001) Rapid changes of glacial climate simulated in a coupled climate model. Nature 409:153–158

Hall M, Bryden i H (1982) Direct estimates and mechanisms of ocean heat transport. Deep-Sea Res 29:339–359

Hansen B, Østerhus S (2000) North Atlantic-Norwegian Sea Exchanges. Prog Oceanogr 45:109–208

Hansen B, Østerhus S, Quadfasel D, Turrell W (2004) Already the day after tomorrow? Science 305:953–954

Hardy JT (2003) Climate change. Causes, effects, solutions. Willey Ltd, p 247

Helland-Hansen B, Nansen F (1909) The Norwegian Sea, its physical oceanography based upon the Norwegian Research 1900-1904. Report on Norwegian and Marine Investigations, vol 2, p 390

Holliday NP, Hughes SL, Bacon S, Beszczynska-Moeller A, Hansen B, Lavın A, Loeng H, Mork KA, Østerhus S, Sherwin T, Walczowski W (2008) Reversal of the 1960s to 1990s freshening trend in the northeast North Atlantic and Nordic Seas. Geophys Res Lett 35:L03614. doi:10.1029/2007GL032675

Holliday NP, Hughes S, Lavin A, Mork KA, Nolan G, Walczowski W, Beszczynska-Moeller A (2007) The end of a trend? The progression of unusually warm and saline water from the eastern North Atlantic into the Arctic Ocean. CLIVAR Exch 121:19–20

Huang RX (1999) Mixing and energetics of the oceanic thermohaline circulation. J Phys Oceanogr 29:727–746

IPCC (2007) Fourth assessment report of the intergovernmental panel on climate change. Cambridge University Press, Cambridge

Knight J, Allan RJ, Folland CK, Vellinga M, Mann ME (2005) Natural variations in the thermohaline circulation and future surface temperature. Geophys Res Lett 32:L20708. doi:1029/2005GL024233

Korty RL, Emanuel KA, Scott J (2008) Tropical cyclone–induced upper-ocean mixing and climate: application to equable climates. J Clim 21, doi: 10.1175/2007JCLI1659.1

Kuhlbrodt T, Griesel A, Montoya M, Levermann A, Hofmann M, Rahmstorf S (2007) On the driving processes of the Atlantic meridional overturning circulation. Reviews of Geophysics 45:RG2001, doi:10.1029/2004RG000166

Loeng H, Brander K, Carmack E, Denisenko S, Drinkwater K, Hansen B, Kovacs K, Livingston P, McLaughlin F, Sakshaug E (2005) Marine Systems. Arctic Clim Impact Assess ACIA, Arctic & Antarctic Research Institute, St. Petersburg, Russia

Luterbacher J, Dietrich D, Xoplaki E, Grosjean M, Wanner H (2004) European Seasonal and Annual Temperature Variability, Trends, and Extremes Since 1500. Science 303:1499–1503

Metcalf WG (1955) On the formation of bottom water in the Norwegian Basin. Transactions. Am Geophy Union 36(4):596–600

Munk W, Wunsch C (1998) Abyssal recipes II: energetics of wind and tidal mixing. Deep-Sea Res 45:1977–2010

Nesje J, Jansen E, Birks H, Bjune A, Bakke J, Andersson C, Dahl S, Kristensen D, Lauritzen S, Lie Ø, Risebrobakken B, Svendsen J (2005) Holocene climate variability in the Northern North Atlantic Region: a review of terrestrial and marine evidence. The Nordic Seas. An Integrated Perspective 158:289–322 AGU Geophysical Monograph

Orsi AH, Smethie WM, Bullister JL (2002) On the total input of Antarctic waters to the deep ocean: a preliminary estimate from chlorofluorocarbon measurements. J Geophys Res 107C8:3122 doi:10.1029/2001JC000976

Overland J, Spillane MC, Soreide NN (2004) Integrated analysis of physical and biological pan-arctic change. Clim Change 63:291–322

Parkinson CL, Cavalieri DJ (2008) Arctic Sea Ice Variability and Trends, 1979–2006. J Geophys Res 113:C07004. doi:10.1029/2007JC004564

Proshutinsky AY, Johnson MA (1997) Two circulation regimes of the wind-driven Arctic Ocean. J Geophys Res 102(C6):12493–12514

Proshutinsky AY, Johnson M (2001) Two regimes of the arctic's circulation from ocean models with ice and contaminants. Mar Pollut Bull 43:61–70

Rahmstorf S (1999b) Currents of change. Investigating the Ocean's role in climate, Essay for the McDonnell foundation centennial fellowship

Rahmstorf S (2000) The thermohaline ocean circulation—a system with dangerous thresholds? Clim Change 46:247–256

Rahmstorf S (2003) The current climate. Nature 421

Rahmstorf S (2006) Thermohaline Ocean Circulation. In: Elias SA (ed) Encyclopaedia of quaternary sciences. Elsevier, Amsterdam

Rahmstorf S, Ganopolski A (1999) Long-term global warming scenarios computed with an efficient couplet climatic model. Clim Change 43:353–367

Richardson PL (2008) On the history of meridional overturning circulation schematic diagrams. Prog Oceanogr 76:466–486

Ronski S, Budeus G (2005) Time series of winter convection in the Greenland Sea. J Geophys Res 110:C04015. doi:10.1029/2004JC002318

Rothrock DA, Percival DB, Wensnahan M (2008) The decline in Arctic sea-ice thickness: separating the spatial, annual and interannual variability in a quarter century of submarine data. J Geophys Res 113:C05003. doi:10.1029/2007JC004252

Schauer U, Beszczynska–Moeller A, Walczowski W, Fahrbach E, Piechura J, Hansen E (2008) Variation of measured heat flow through the fram strait between 1997 and 2006. In: Arctic-subarctic ocean fluxes, Springer Science, pp 15–43

Schiermeier Q (2006) A sea change. Nature 439:256–260

Serreze MC, Francis J (2006) The Arctic amplification debate. Clim Change. Springer, doi: 10.1007/s10584-005-9017-y

Stewart RH (2006) Introduction to physical oceanography. Department of Oceanography, Texas A & M University

Steynor A, Wallace C (2007) The Gulf Stream—Atlantic meridional overturning circulation: observations and projections. UK Climate Impacts Programme

Stommel H (1961) Thermohaline convection with two stable regimes of flow. Tellus 13:131–149

Stouffer RJ, Yin J, Gregory JM, Dixon K, Spelman M, Hurlin W, Weaver A, Eby M, Flato G, Hasumi H, Hu A, Jungclaus J, Kamenkovich I, Levermann A, Montoya M, Murakami S, Nawrath S, Oka A, Pelitie WR, Robitaille DY, Sokolov A, Vettoretti G, Webber SL (2006) Investigating the causes of the response of the thermohaline circulation to past and future climate changes. J Clim 19:1365–1387

Talley LD (2003) Shallow, intermediate and deep overturning components of the global heat budget. J Phys Oceanogr 33:530–560

Talley LD, Reid JL, Robbins PE (2003) Data-based meridional overturning stream functions for the global ocean. J Clim 16:3213–3226

Toggweiler JR, Samuels B (1995) Effect of Drake Passage on the global thermohaline circulation. Deep-Sea Res 42:477–500

Wunsch C (2002) What is the thermohaline circulation? Science 298:1179

Wunsch C, Ferrari R (2004) Vertical mixing, energy and the general circulation of the oceans. Annu Rev Fluid Mech 36:281–314

Chapter 2
The Nordic Seas

The Nordic Seas is a traditional name of the Greenland Sea, the Iceland Sea, and the Norwegian Sea. Some authors include the Barents Sea in the Nordic Seas, as well. In the 1980s, American oceanographic literature made unsuccessful attempts to call them "The GIN Seas" (from the first letters of the seas). Although they cover only a small part of the world ocean, the Nordic Seas play unusual role in its circulation and the dynamics of the northern hemisphere's climate. The importance of the Nordic Seas is in their location. Together with the Arctic Ocean, the NS make up the Arctic Mediterranean Sea, extending from the Greenland-Scotland Ridge to the Bering Strait. With respect to the bathmetry, it is the northernmost part of the Atlantic Ocean. Being opened from the south to the North Atlantic and from the north to the Arctic Ocean, connected on the east with the shelf Barents Sea and indirectly also with the Arctic Ocean, the NS form a melting pot in which various water masses mix and transform. It is also a place where the most world ocean's deep and intermediate water is formed. The processes make the Nordic Seas an area of the most positive air temperature deviations from the zonal average (Rahmstorf and Ganopolski 1999). Location in the polar zone and concurrent massive heat inflow from the south causes the region to have a tendency of natural climate changes and to be extremely susceptible to external factors, both natural and anthropogenic.

2.1 Characteristics of the Nordic Seas

The Nordic Seas cover the area north of the Greenland-Scotland Ridge, south of the Fram Strait, bordered on the west by the Greenland shore and on the east by a conventionally assumed line between the south Spitsbergen and the north Norway. They cover approximately 2.5×10^6 km^2—0.75 % of the world ocean's area. Despite such a small area, it is an extremely diversified and dynamic sea region:

W. Walczowski, *Atlantic Water in the Nordic Seas*,
GeoPlanet: Earth and Planetary Sciences, DOI: 10.1007/978-3-319-01279-7_2,

- The bottom topography is complex; there are shallow and wide shelves with steep slopes as well as deep ocean basins intersected by underwater ridges;
- The dynamic scale of processes (Rossby radius of deformation) is small—from a few to several kilometres;
- The interactions between the atmosphere and the ocean are strong. The exchange of momentum, heat, gases and water is large particularly in winter, between November and April;
- Cold and fresh waters from high latitudes come across warm, saline waters from the south and interact in various ways: by mixing along the hydrographic fronts, deep convection, subduction, entrainment;
- Sea ice is formed in winter on the north and west edges, in summer the Nordic Seas are virtually free of ice;
- Carbon flux from the atmosphere to the ocean (per area unit) is one of the highest in the world ocean;
- The primary production exceeds the world ocean's average primary production, there are rich fish resources, particularly in the area which is under the influence of warm and nutrient-rich Atlantic water.

All the above mentioned properties of the Nordic Seas make the region exceptionally significant in a broadly defined climate context:

- About 6 Sv of cold deep water spills southward over the Greenland-Scotland Ridge. It is one-third of the volume transported by the AMOC, however, after mixing and entrainment of ambient water while flowing down the southern slope of the ridge, it drives about two-thirds of the AMOC volume transport (Hansen et al. 2004);
- Intense surface forcing in the Nordic Seas has great impact on the hydrographic properties of the water flowing into the Atlantic Ocean, both as deep overflow above the Greenland-Scotland Ridge and surface transport through the Denmark Strait (Hansen and Østerhus 2000).
- The Atlantic Water is also strongly modified by the surface exchange before its potential arrival into the Arctic Ocean (Karcher et al. 2003);
- Freshwater transport through the Nordic Seas is significant. Besides the Canadian Arctic Archipelago, the Nordic Seas are an essential route for the hydrographic cycle of the Arctic (Aagaard and Carmack 1989);
- The atmospheric circulation of the northern hemisphere is sensitive to sea ice distribution and sea surface temperature in the Nordic Seas, the Labrador Sea, the Barents Sea and the Kara Sea (Deser et al. 2004);
- The Nordic Seas is one of few regions of the world ocean which absorbs large quantities of the major greenhouse gas—carbon dioxide (CO_2). Annual atmosphere-to-ocean carbon flux is one of the highest, oscillating between 20 and 85 g m^{-2} (Anderson et al. 2000).

2.2 Climate

Unusually mild climate has had tremendous importance for the survival of people inhabiting the shores of the Nordic Seas for ages. Prevailing south-westerly winds, warm ocean currents and the heat accumulated in summer in the water column make the mean air temperature in the Nordic Seas region 10–20 °C higher than the average for such latitudes elsewhere (Drange et al. 2005). Due to the system of sea currents, sea ice along the western coast of Greenland extends far to the south, whereas the Norwegian coastline and the areas farther north are free of ice, even in winter. The large-scale atmospheric circulation variability in the region is mostly influenced by the NAO (Furevik and Nilsen 2005). The wind field is strongly related to the Icelandic Low and the route of western storms travelling from the North America. In the regional scale, the Norwegian Sea and the Barents Sea are significant sources of winds where mesoscale polar lows may form at the ice margin in winter.

The combined large volume of heat imported from the south and location in the polar zone makes the region very sensitive to any climate change, including those forced internally. In the period of intensified westerly winds, a great deal of primary oceanic indices shows unprecedented growing tendency. Climate models indicate that this may be related to the increasing contents of greenhouse gases in the atmosphere.

2.3 Bathymetry

The main bathymetric characteristics of the Nordic Seas are shown in Fig. 2.1 drawn up on the basis of bathymetric data from ETOPO2, 2 min grid (US 2006). On the south, the Greenland-Scotland Ridge, with the deepest trenches ($\sim$850 m) located at the Faroe-Shetland Channel and the Denmark Strait ($\sim$620 m), creates a natural border with the North Atlantic. The Fram Strait, with a sill at the depth of 2,600 m, is the main connection with the Arctic Ocean. The most important connection with the Barents Sea is the Bear Island Trough, situated between the north Norway and the Bear Island. The region is frequently referred to as the Barents Sea Opening (BSO).

The bottom formations within the Nordic Seas divide the sea region into the Greenland Sea, the Iceland Sea and the Norwegian Sea. The most important of these include the following oceanic ridges: the Kolbeinsey Ridge, the Mohn Ridge and the Knipovich Ridge. The Kolbeinsey Ridge stretches northwards from the shelf north of Iceland to Jan Mayen Island, where it is intersected by the Jan Mayen Fracture Zone. The Mohn Ridge starts at Jan Mayen and extends towards the north-east. Its depth is between 1,000 and 2,000 m and it is rather a group of isolated elevations, underwater mountains, than a homogenous ridge. The Knipovich Ridge, stretching from the Mohn Ridge to the Fram Strait, is of similar

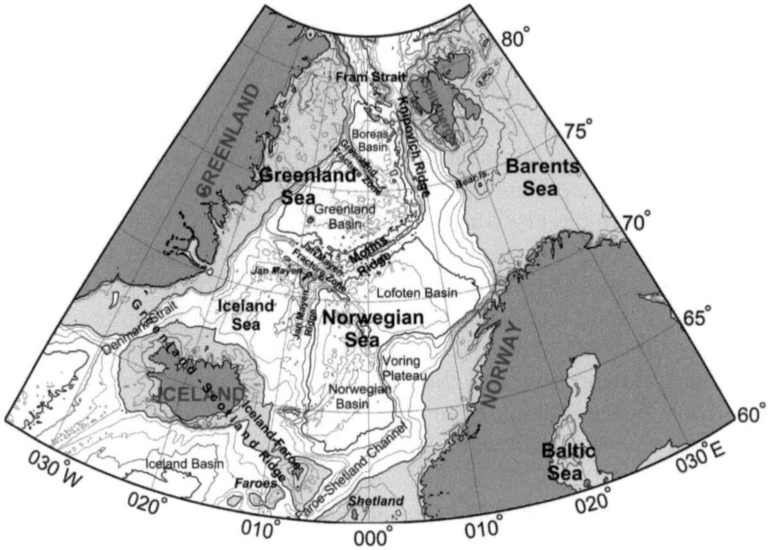

Fig. 2.1 Bathymetry of the Nordic Seas based on ETOPO2 data (NOAA 2006). The 500 and 2,500 m isobaths are thickened

structure. The highest peaks project more than 1,000 m above the ridges, which is of great importance for the generation of currents in that area, including in particular the eddies generated by topography (Walczowski 1997).

The seamounts divide the Nordic Seas into 4 major basins. To the west of the Mohn and Knipovich Ridge system there are two Greenland Sea basins: the larger Greenland Basin, reaching depths from 3,400 to 3,600 m, divided on the north by the Greenland Fracture Zone from the shallower (of a depth of ca. 3,200 m) Boreas Basin. The Norwegian Sea also has two large basins: the Norwegian and the Lofoten Basin. The first of those is the largest and deepest basin of the Nordic Seas, extending to the north of the Greenland-Scotland Ridge, to the east of the Iceland Plateau, towards the Vøring Plateau and the continental slope of Norway. The average depth of the Norwegian Basin is between 3,200 and 3,600 m with extremes of 3,800 m. To the north of the Greenland Basin lies the smaller Lofoten Basin with the average depth of 3,200 m.

2.4 Ocean Currents

The Arctic Ocean with the adjacent seas is a mediterranean type of sea, for the link and communication with the main basins of the world ocean is limited (Dietrich et al. 1980). Due to that the circulation in the Arctic is mainly driven by thermohaline mechanisms and only modified by wind (Tomczak and Godrey 1994).

The inflow of water from the North Atlantic into the Nordic Sea occurs only in three regions (Fig. 2.2). Due to complex and incoherent terminology, only the alternative names of currents are given in the figure. The North Icelandic Current, also referred to as the Icelandic branch (Hansen et al. 2008) flows to the west of Iceland, through the eastern part of the Denmark Strait. The Faroe Current flows over the Iceland-Faroe Ridge. The third current, also known as the Shetland Current, flows between the Faroe Islands and the Shetlands through the Faroe-Shetland Channel. Despite the similar origin, the properties of the waters are different (Table 2.1) (Blindheim and Østerhus 2005; Hansen et al. 2008).

The estimated volume transport by individual branches was given by many authors, the total Atlantic Water transport was provided mainly on the basis of calculations and it varies from a few to several Sverdrups. According to recent estimates, the total volume transport of the Atlantic Water into the Nordic Seas is between 7.7 and 9 Sv (Hansen et al. 2008), and the heat flux is between 276 and 310 TW. The Faroe Current carries most of water, and due to higher temperature, more heat flows through the Faroe-Shetland Channel.

The main current flowing in the eastern area of the Denmark Strait is the North Icelandic Current. It if formed as a result of splitting of the Irminger Current flowing from the south which carries the Atlantic Water originating from the North Atlantic Current. Having passed Iceland, the current turns east and mixes with the Arctic water carried by the East Icelandic Current (EIC). Jónsson and Briem

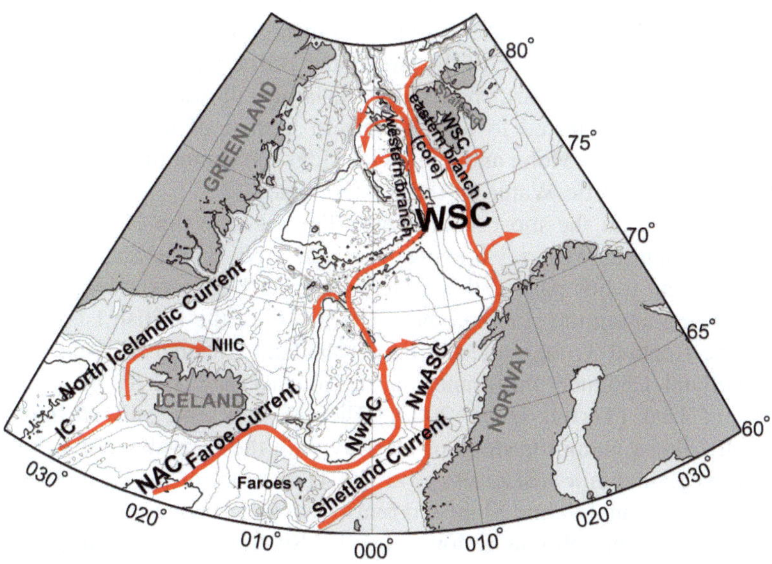

Fig. 2.2 The main routes of the Atlantic Water in the Nordic Seas. *NIIC* North Icelandic Irminger Current, *NAC* North Atlantic Current, *NwAC* Norwegian-Atlantic Current, *NwASC* Norwegian Atlantic Slope Current, *WSC* West Spitsbergen Current

Table 2.1 Temperature (T), salinity (S), volume transport (Vol) and heat transport (Q), carried by the main branches of the Atlantic Water into the Nordic Seas from the south

Current	Water	Origin	T (°C)	S	Vol (Sv)	Q (TW)
North Icelandic	Irminger Sea Water (ISW)	North Atlantic Current	5–7	35.05–35.10	1	25
Faroe	Modified North Atlantic Water (MNAW)	North Atlantic Current	7–8.5	35.1–35.3	3.5	124
Shetland	Northeast Atlantic Water (NEAW)	The current along continental slope to the west of the British Isles	9.5	35.32	3.2	127

(2003) estimate the average volume transport of the North Icelandic Current at 1 Sv and the heat flux at 25 TW.

The flow over the Faroe-Iceland Ridge (the Faroe Current) transports the Modified North Atlantic Water (MNAW). In that area the current carries 3.5 Sv of water, and 124 TW of heat (Hansen et al. 2003). Having crossed the Faroe-Iceland Ridge, it merges with the Arctic water of the East Iceland Current to form the Iceland-Faroe Front, turns and flows eastward over the northern slope of the ridge. In this work it is referred to, after Orvik and Niiler (2002), as the Norwegian-Atlantic Current (NwAC). A portion of the water recirculates towards the Faroe-Shetland Channel, whereas the major part flows to the Norwegian Basin, approaches the Vøring Plateau, and turns north-west towards the Jan Mayen. In the vicinity of Jan Mayen, the current divides into three branches: a part of it recirculates along the western edge of the Greenland Basin, another part flows into the Iceland Sea, and the third part, presumably the largest, flows to the north-east. That branch, hereinafter referred to as the western branch of the West Spitsbergen Current, flows over the Mohn and Knipovich ridges, forming the Arctic Front (AF) together with the Arctic Water on the western side.

The warmest and the most saline water flowing into the Nordic Seas is the Northeast Atlantic Water, flowing through the Faroe-Shetland Channel (Table 2.1). The major part of that water is carried by a topographically-steered current flowing along the slope on the western side of the British Isles, and flows through the Rockall Trough and the Faroe-Shetland Channel. Having passed the Faroe-Shetland Channel, the branch carries on average 3.2 Sv of volume and 127 TW of heat (Turrell et al. 2003). Then the current, still steered by the topography, flows along the slope of Norway with the core above the shelf edge. After Norwegian authors (Skagseth et al. 2004) that part of the current is referred to as the Norwegian Atlantic Slope Current (NwASC) Having passed the northern Norway, the current splits as follows: 1.3–1.7 Sv (Ingvaldsen et al. 2004) flows eastward to the Barents Sea through the Bear Island Trough and the remaining part flows northwards along the Barents Sea slope and the west Spitsbergen shelf as the West Spitsbergen Current.

The frontal zone separating the Atlantic Water from the Arctic Water extends above the system of underwater ridges of the Norwegian Sea through the Greenland Sea. These include, respectively from the south: the Jan Mayen Ridge, the Mohn Ridge and the Knipovich Ridge. The hydrographic front which divides the both domains has different names, which is sometimes troublesome. The majority of authors call it the Arctic Front, however, some refer to it as the Polar Front, the Arctic Frontal Zone or the Polar Ocean Front. The author of the present work consistently uses the name Arctic Front after Swift (1986) in order to distinguish it from the Polar Front which borders the Atlantic Domain from the east. The western branch of the WSC is made of baroclinic currents formed at the front. A part of the Atlantic Water carried northward by the western branch of the WSC crosses the Arctic Front as eddies and flows into the Greenland Basin (Piechura and Walczowski 1995; Walczowski 1997), and a part recirculates near Spitsbergen, between 76°N and 78°N latitudes, into the Boreas Basin. At 78°N, due to the bottom topography the both branches of the WSC converge (Walczowski et al. 2005). Further up north, the current divides again to form a complex, multi-branch structure in the Fram Strait (Manley 1995).

The most AW from the western branch recirculates westward in the Fram Strait, towards Greenland, and then flows to the south with the East Greenland Current as the Return Atlantic Water (RAW). The major part of the water carried by the eastern branch of the WSC (core) flows into the Arctic Ocean as the so-called Svalbard Branch (SB), and then circulates cyclonically around the Arctic Ocean shelf slope (Rudels et al. 1999), and some portion of it flows out on the western side of the Fram Strait.

The exchange in the Fram Strait, at 79°N latitude, is of approximately 12 Sv to the north and 14 Sv to the south, with the resulting average southward flow of 2 Sv (Schauer et al. 2008). More than a half of that transport takes place in the layer deeper than 700 m, which is below the 0 °C isotherm. The main northward flow towards the Arctic Ocean occurs in the eastern part of the Fram Strait by the West Spitsbergen Current. Schauer et al. (2004) estimate the value of transport as 3–5 Sv, with high monthly and annual variability. Walczowski et al. (2005) state measured value of 3–6 Sv and emphasize high variability of flow on the scale of days. The East Greenland Current, carrying approximately 3–4 Sv of water, including 1 Sv of the Polar Water, flows from the Arctic Ocean through the Fram Strait.

In spite of the involvement of international projects and the growing density of the mooring array measuring water temperature, velocity and direction of currents, the Fram Strait balance remains inaccurate. The most recent observations indicate the net average southward volume flow of approximately 2.5 Sv and ca. 30 TW northward heat flux (Schauer et al. 2008).

The East Greenland Current, which is steered by the bottom topography flows along the Greenland continental margin, carries different water masses and sea ice southwards. In addition to the Return Atlantic Water and the Polar Water, the EGC carries the intermediate and deep water from the Arctic Ocean (Rudels et al. 2002), as well as the Pacific Water. Due to the bottom topography, a significant part of the flow above 1,600 m and the entire flow below that depth turns east as the Jan Mayen Current (JMC), thus forming the southern branch of the cyclonic circulation around the Greenland Basin, called the Greenland Sea Gyre (GSG). Further south, a part of the EGC surface waters joins the East Iceland Current and flows eastwards. The remaining part of the EGC leaves the Nordic Seas through the Denmark Strait. The volume transport in that part of the EGC is estimated at 1.3 Sv (Blindheim and Østerhus 2005).

2.5 Two Branches of the West Spitsbergen Current

The first map of circulation in the Nordic Seas was drawn up by Mohn (Fig. 2.3) and presented in a report by Helland-Hansen and Nansen (1909). According to that interpretation, most of the AW flows into the Norwegian Sea between the Shetlands and the coast of Norway, whereas the minor part flows through the Faroe-Shetland Channel and a negligible part to the west of the Faroe Islands. The branches then merge into a broad current flowing north-east of the Norwegian coast. Having passed the North Cap, the current turns to the east and flows into the Barents Sea where it recirculates back to the Norwegian Sea and the Greenland Sea. A part of the AW flows from the Barents Sea into the Arctic Ocean, and a small portion flows through the Fram Strait along the coast of Spitsbergen. An intensive current from the Arctic Ocean flows on the western side of the Fram Strait. The central part of the Norwegian Sea is occupied by a cyclonic eddy.

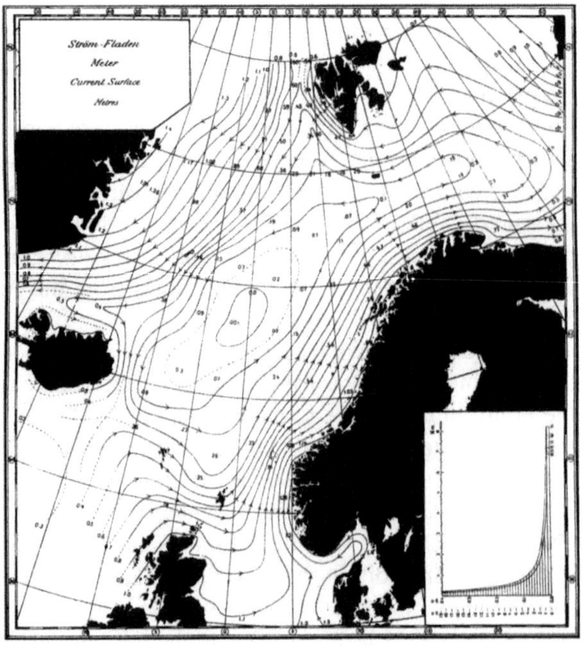

Fig. 2.3 The circulation of the Nordic Seas according to Mohn (After Helland-Hansen and Nansen 1909)

The same paper includes a diagram of the currents according to the authors (Fig. 2.4). In that version, the AW flows into the Norwegian Sea mainly through the Faroe-Shetland Channel, and a small portion of it flows as eddies west of the Faroe Islands. The AW flows along the shelf of Norway and then divides into two branches: the eastern one, flowing into the Barents Sea and the northern one which flows along the shelf towards the Fram Strait. West of Spitsbergen the major part of the AW recirculates and a part of it flows into the Arctic Ocean along the Spitsbergen shelf. A broad current flows from the Arctic Ocean along the eastern Greenland. There are large cyclonic eddies in central parts of the Norwegian, Greenland and Barents Seas.

It is amazing that the authors of those maps were able to reconstruct the system of currents in the studied areas with a relatively high accuracy. They show most of the existing and known currents, including such a detail as the Baltic Sea origin of the current flowing nearest to the coast of Norway—the Norwegian Coastal Current (NCC). No wonder such an image of flows has dominated the comprehension of circulation in the Nordic Seas region for decades.

Traditionally, a lot of papers describe the West Spitsbergen Current as a simple barotropic flow related to the continental slope of the Barents Sea and Spitsbergen. And yet it is visible from the above figure alone that the flow is much more complex, made up of several streams with particularly visible zone of baroclinicity and the related flow over the Mohn and Knipovich Ridges. The existence of the second, western branch of the WSC has been often claimed. Old Soviet maps (Aleseev and Istoshin 1956) show much more complex structures, and in 1973 Wegner presented an interesting chart of currents with specified transports values (Fig. 2.5).

Fig. 2.4 The circulation of the Nordic Seas according to the observations of Helland-Hansen and Nansen (1909)

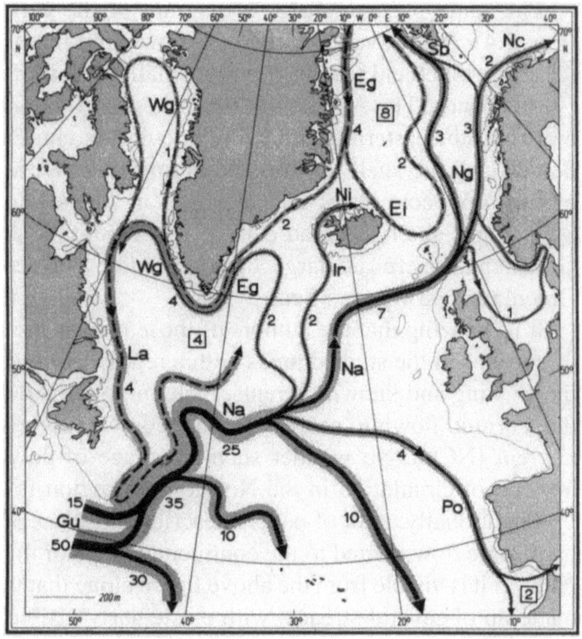

Fig. 2.5 Scheme showing the transport of water in the North Atlantic according to Wegner (1973), after Dietrich et al. (1980)

The map shows the North Atlantic Current branching off the Gulf Stream which initially carries 25 Sv to deliver, after further branching, 2 Sv to the Iceland Sea with the Irminger Current and 7 Sv to the Norwegian Sea through the Faroe-Shetland Channel. According to that interpretation, having passed the Faroe-Shetland Channel, the current (herein referred to as the Norwegian Current) divides into a branch which flows along the coast and the western branch which probably flows above the underwater ridges. Each of them transports 3 Sv of water. The branch which flows along the coastline, having passed Norway, divides again, and the majority of water flows into the Barents Sea, whereas 1 Sv flows along the shelf edge of the Barents Sea. That part of the flow is most frequently called the West Spitsbergen Current. According to Wegner, the western branch joins the eastern one at 76°N and then continues its flow as the Spitsbergen Current, carrying a total of 4 Sv of the Atlantic Water.

The IOPAS oceanographers also found baroclinic currents in that area from observations in the Greenland Sea and crossing the Arctic Front. The program of studying mesoscale structures above the Arctic Front yielded a lot of results and developed knowledge about flows in that region. In the PhD thesis (Walczowski 1997), developed as a result of that process-oriented study, the author provided a map with two branches of the Atlantic Water converging at 78°N latitude (Fig. 2.6). That was previously mentioned by Piechura and Walczowski (1995).

However, when an attempt was made to present the views at the first meeting of the international VEINS program, they were not favourably received. Nevertheless, the VEINS program made it possible to gain deeper comprehension of the

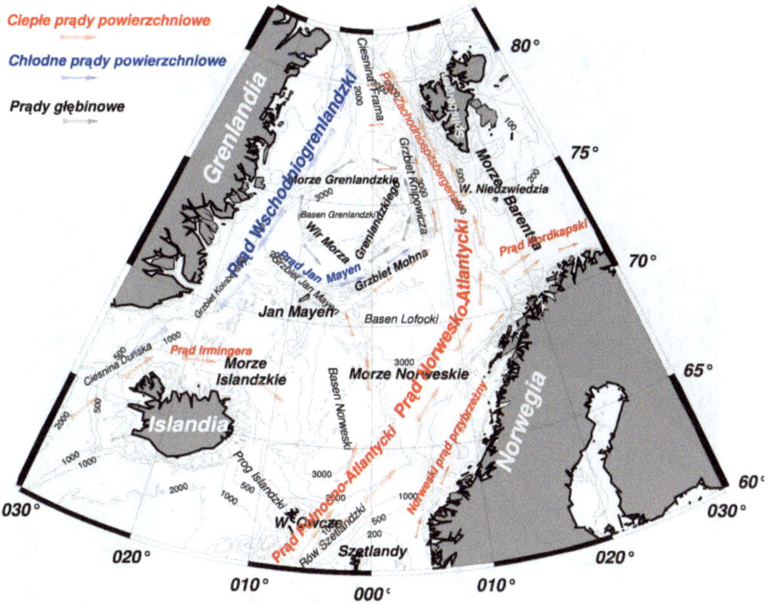

Fig. 2.6 The circulation of the Nordic Seas (Walczowski 1997—Ph. D. thesis)

structure of the AW inflow into the Nordic Seas. In spite of the fact that the AW transport over the Faroe-Iceland Ridge had been previously described by Hallend-Hansen and Nansen (1909), a valid paradigm was the opinion that most of the AW flows into the Nordic Seas through the Faroe-Shetland Channel, and an insignificant part is transported by the Irminger Current and the North Iceland Current between Greenland and Iceland. That is the reason why the maps showing two WSC branches (including the map placed in the author's PhD thesis) showed the western WSC branch being a result of splitting of the current inflowing through the Faroe-Shetland Channel. It was not until 1996 (Poulain et al. 1996) that a paper was published describing two branches of the Norwegian-Atlantic Current in the Norwegian Sea, and in 2000, studies by Hansen and Østerhus showed that a major part of the AW flows into the Norwegian Sea over the Greenland-Scotland Ridge, between Iceland and the Faroes. This enabled to draw conclusions as to the origin of the western WSC branch. However, even the above-mentioned authors indicate two AW branches only in the Norwegian Sea, not extending them into the Greenland Sea.

The idea of two WSC branches was spread by the IOPAS oceanographers with great perseverance in the published papers (Piechura and Walczowski 1995; Walczowski 1997, the VEINS report, Walczowski et al. 2005; Walczowski and Piechura 2006; Walczowski and Piechura 2007), at numerous international conferences, including in particular the annual European Geosciences Union assemblies, as well as in the reports under the ASOF-N and DAMOCLES programs.

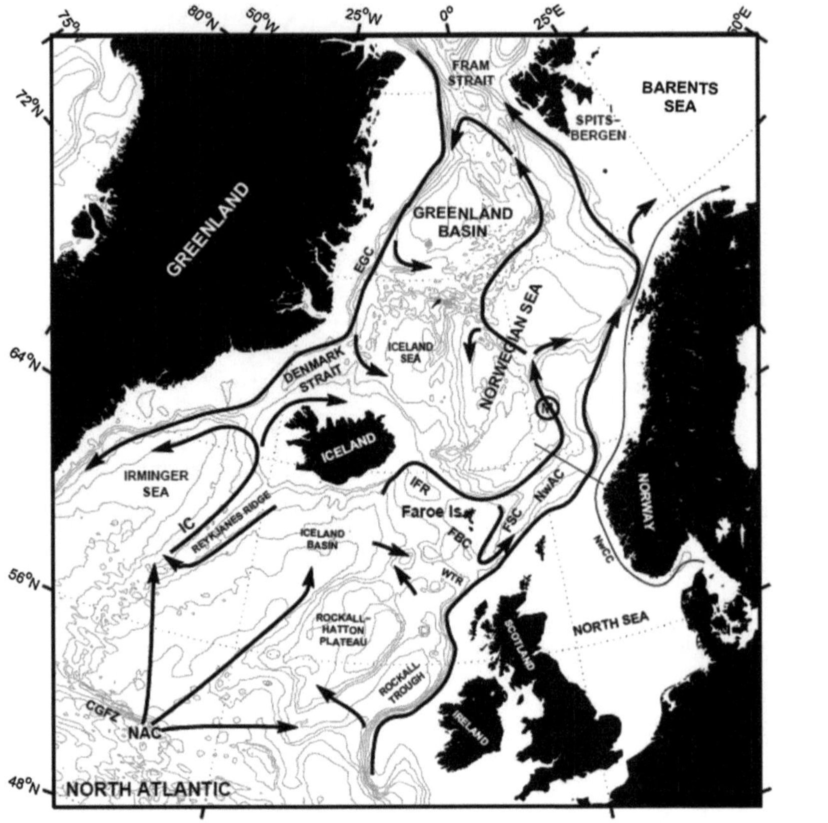

Fig. 2.7 Surface current regime in the North Atlantic and Nordic Seas (after Furevik and Nielsen 2005)

However, the first to obtain some wide publicity was a paper published by Orvik and Niiler in (2002). Nevertheless, a great number of contemporary papers base on the old, single-branch WSC pattern. The example of modern vision of the sea currents in the North Atlantic region is the pattern included in the paper by Furevik and Nielsen (2005) (Fig. 2.7).

2.6 Ocean–Atmosphere Interaction

The inflow of the Atlantic Water into the Nordic Seas is one of key features to shape local climate and ecosystem of the region; it is also essential for the climate change of the northern hemisphere. Predicting the AW inflow into the region in a time scale ranging from months to years is thus essential to understand the climate-shaping processes and to make economic and political decisions.

In 2003 Orvik and Skagseth demonstrated that the Atlantic Water volume carried by the NwASC through the Svinoy hydrographic section at approximate latitude of 62°N responds to changes of wind above the North Atlantic at 55°N latitude with a delay of about 1.5 year orvik and skagseth 2003. It proved that the main AW inflow branch may be predictable to some extent and the inflow is affected by the wind field.

The pattern of winter winds above the northern hemisphere is shaped by high-pressure systems in the subtropical zones and low-pressure systems above the subpolar Atlantic and Pacific. In the case of the North Atlantic, the wind pattern is shaped mostly by the Azores High and the Icelandic Low. The changing arrangement of the above-mentioned atmospheric systems in relation to each other is a leading driver of the wind field variability in that region. The variability is described by the North Atlantic Oscillation (NAO) (Hurrell 1995, Fig. 2.8). The NAO index is a standardized difference of atmospheric pressure at the sea level in Lisbon and Reykjavik. Winter NAO index, calculated for the December-March period, is of particular significance. The positive phase of the NAO occurs during a large-scale shift of the air masses from the Icelandic Low, the Nordic Seas and the Arctic Ocean to the Azores High. Strong Azores High and deep Icelandic Low then generate strong gradient of atmospheric pressure and thus intensify the westerly winds. That results in higher temperature and precipitation over the northern Europe, and lower temperature and precipitation over the southern Europe. The winds and their variability have therefore a strong impact upon climate in Europe, as well as on a broad range of physical, ecological and socio-economic conditions of the Atlantic region. The causes of the variable location of the atmospheric systems and the NAO occurrence are not entirely recognized and their analysis is beyond the scope of the present work.

The mechanisms of the impact of the atmospheric circulation, including NAO, upon the ocean hydrography and circulation vary, as well as the respective time scales which range from days to years. The fastest reaction in dynamics occurs as a result of the change in wind friction on the sea surface and the Ekman drift which changes the Sea Surface Height (SSH). That generates barotropic currents forced

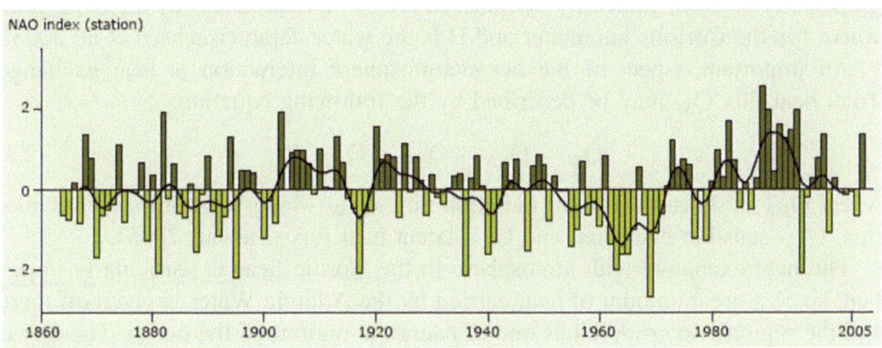

Fig. 2.8 Winter NAO index from 1864 to 2007 (EEA 2008)

by horizontal pressure gradient (Druet and Kowalik 1970). Walczowski et al. (2005) observed that the described mechanism caused short-term (on the scale of days) fluctuations of the West Spitsbergen Current. As far as hydrography is concerned, it causes a change in the sea surface temperature and the depth of mixing layer. Slower, baroclinic reaction occurs when the water density structure changes as a result of anomaly of driving forces: wind field, heat exchange with atmosphere or fresh water supply. The mechanisms may have local impact or affect the areas where water masses form. In the latter case, the signal is advected as an anomaly together with mean flow, or propagates as a slow baroclinic wave (Furevik and Nilsen 2005).

The Gulf Stream and its northern extension—the North Atlantic Current—are also sensible to atmospheric forcing on a wide spectrum of time scales. Also the two most important AW branches flowing into the Nordic Seas seem to be sensible to changes in the large-scale atmospheric circulation (Furevik et al. 2007). Data gathered by drifters, measurements of temperature and sea surface height indicate that in years of positive NAO, the eastern branch flowing into the Nordic Seas via the Faroe-Shetland Channel intensifies (Flatau et al. 2003). That was confirmed by the study of the NwASC described by Orvik and Skagseth (2003) at the Svinoy section. 15-months' time lag between wind curl at 55°N latitude and flows at 62°N may indicate the baroclinic nature of the reaction. In that particular case, the described mechanism was more complicated—it was of baroclinic-barotropic nature. Other works also indicate the baroclinic reaction to a change in driving forces. Blindheim et al. (2000) show that the meridional extent of the AW in the south-eastern region of the Nordic Seas is correlated with the NAO index with a 3 years' time lag. They explain that by a change in the AW inflow between Iceland and Scotland. Other studies (Mork and Blindheim 2000) show that baroclinic transports in the eastern and western branch of the NAC are in opposite phases. However, precise explanation of mechanisms relating the NAO to the flows is not available.

It is also not clearly known which of the drivers—the inertia of inflowing currents, local winds, or buoyant forces—dominates in forcing the AW flows in the Nordic Seas. It has been known for long that currents in the region are topographically-steered, guided by isobaths, and to be more precise—by the f/H isoline, where f is the Coriolis parameter and H is the water depth (Isachsen et al. 2003).

An important aspect of the ocean/atmosphere interaction is heat exchange. Total heat flux Q_W may be described by the following equation:

$$Q_W = Q_{SW} + Q_{LW} + Q_S + Q_L \tag{2.1}$$

where Q_{SW} is short-wave solar radiation flux, Q_{LW}—long-wave infrared radiation flux, Q_S—sensible heat flux, and Q_L—latent heat flux (Stewart 2006).

The heat exchange with atmosphere in the Nordic Seas is particularly important, since a great amount of heat carried by the Atlantic Water is given off there, and the region is considered as one of energetic regions of the ocean. The flux of latent heat absorbed or released during water phase transitions, is the main component of the heat budget. It's typical values for the AW inflow area range between 80 and 100 W/m^2. The second component is sensible heat flux, which is between 40 and 80 W/m^2 there, with a maximum of up to 100 W/m^2 in the marginal ice zone. For most of the Nordic Seas region, all the ocean heat budget components yield an average ocean-atmosphere flux of 160–180 W/m^2, with maxima reaching 300 W/m^2 in the Fram Strait (Furevik and Nielsen 2005). To compare, average daily solar radiation absorbed by the ocean in the tropics equals 200 W/m^2 (Massel 2007). Simonsen and Haugan (1996) estimate that the European subarctic seas release on average 242 TW of heat, of which 136 TW is released by the Barents Sea. The Arctic Ocean releases on average 86 TW of heat to the atmosphere.

The wind field has an important impact on heat exchange between the ocean and atmosphere; the increase of wind speed intensifies the flux of latent and sensible heat from water to atmosphere. High NAO index is reflected by 20 % greater heat release in the north-western Atlantic. In the Nordic Seas the effect is reverse, it is estimated that the ocean-to-atmosphere heat loss is then reduced by 5–10 W/m^2 (Furevik anf Nielsen 2005).

The strength of westerly winds has changed in recent decades, from very weak in the 1960s, to unusually strong in the 1990s. Concurrently, the NAO index changed (Fig. 2.8). The changes could be felt even in summer; the IOPAS cruises in the 1990s always encountered harsh weather conditions. In the beginning of the 21st century the weather was favourable for several years in a row, and the NAO index was low. Increase of the winter NAO index at the turn of 2006/2007 was also felt in the summer of 2007, when the weather resembled the conditions of the 1990s. At that time the Nordic Seas experienced unprecedented changes which had never been seen in the period of instrumental observations of that sea region. Changes of the atmospheric forcing drove stronger AW inflow through the western branch and weaker transport over the Greenland-Scotland Ridge. Changes also occurred in water masses, both the surface and the deep. The changes resulted in a significant increase of temperature and salinity of the Atlantic Water flowing into the Barents Sea and the Fram Strait, increase of the freshwater supply, and reduction of ice cover area in the Arctic Ocean. A change occurred in the convection taking place in the Greenland Sea, which currently produces much less deep water and does not reach to such depths as in the 1960s. This in turn results in reduced deep flow over the Greenland-Scotland Ridge and indicates symptoms of slowing AMOC.

2.7 Climatological Fields of Temperature and Salinity

Before presenting the IOPAS research results, the author decided to show the climatological fields of temperature and salinity for the region. These are averaged long-term measurements with low spatial resolution, both horizontal and vertical. Their advantage is that they cover a period of the entire year. PHC 3.0 data set from the Global Ocean Hydrography (Steel et al. 2001) was used, monthly fields

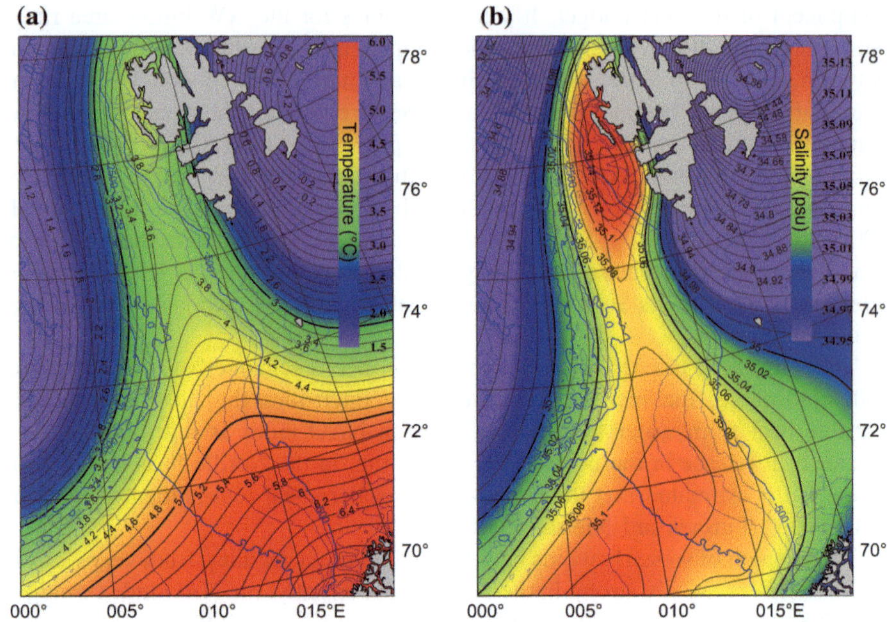

Fig. 2.9 Temperature (**a**) and salinity (**b**) of the Norwegian and Greenland Seas based on PHC 3.0 climatological data for July at depth 100 m

from July, depth of 100 m. The resolution of the PHC climatology is 1° × 1°. The presented horizontal distributions (Fig. 2.9) were obtained by a projection of temperature and salinity data on the cartesian grid used in the case of all other maps included in the work, and then, interpolation of the data with methods applied in other cases (Fig. 5.2). The results show the Atlantic Domain as a wedge-shaped area, wide on the south and tapered towards the Fram Strait, of salinity and temperature higher than in the surrounding regions. The Atlantic water is separated from the Polar and Arctic water masses by hydrographic fronts—areas of high salinity and temperature gradients. The temperature field inside the Atlantic Domain clearly shows the AW transformation: strong meridional temperature gradient. The fields of temperature and salinity show the direction of advection: to the north into the Arctic Ocean and to the west of the Barents Sea.

2.8 Transformation of the Atlantic Water

A strong seasonal signal is contained in the temperature of the AW flowing into the Nordic Seas. Since it is an advective signal, it does not have to correspond to a season, e.g. at the Svinoy section, the AW temperature maximum is in December and the amplitude reaches 0.5 °C (Orvik and Skagseth 2005). The seasonal signal is not vertically homogenous either; the phase is delayed and the amplitude decreases with depth. Multi-year changes of temperature contribute to the seasonal signal. Both signals—seasonal and multi-year—overlap and are carried along with the northward and eastward advection of water masses. During that, the AW undergoes a significant transformation—it mixes with the surrounding waters, cools down due to the interaction with atmosphere. On its way between the Svinoy section and the Gimsoy section, which crosses the Lofoten 400 km up north, the AW carried by the NwASC cools down by an average of 1 °C (from 7–8 to 6–7 °C) and salinity is reduced from 35.20 to 35.15. Having passed the northern coast of Norway, the processes become much more intensive. In the branch that turns eastward, the AW quickly loses its properties in the shallow Barents Sea, cools down and mixes with the surrounding Polar Waters (Schauer et al. 2002). Also the AW flowing towards the Fram Strait is subjected to a very strong transformation, particularly cooling. Between 70°N and 79°N latitude the AW core at the depth of 200 m cools down from 6 to 2 °C on average, and its salinity decreases from 35.14 to 34.95 (Furevik et al. 2002). This process is not uniform, neither in time nor in space. The average annual ocean-to-atmosphere heat flux in the area between Norway and the Fram Strait is 100 W/m^2, however, in winter average heat fluxes in the Fram Strait reach as much as 300 W/m^2 (NCEP/NCAR data, Kelnay et al. 1996). Strong mixing, cooling and decrease of salinity occur at the boundary of the AW with other water masses. Despite such considerable changes, the AW propagates the signal from the source region to the Arctic Ocean. Usually, it is a rise or a fall of temperature. After such a long way and strong transformation, the signal is modified, yet still recognizable, particularly when anomalies are used instead of absolute values. However, the AW temperature anomalies may also be of local origin. They may be caused by a change in the conditions of the ocean–atmosphere exchange or by some dynamic processes. The anticyclonic eddy has a tendency to reduce the AW level, lower the temperature isoline, and, consequently, increase the temperature at a given level. Such an eddy may, due to its depth, carry a considerable heat surplus.

References

Aagaard K, Carmack E (1989) The role of sea ice and other fresh water in the arctic circulation. J Geophys Res 94:14485–14498

Aleseev A, Istoshin B (1956) Chart of constant currents in the Norwegian and Greenland Seas. Trans Knivovich Pol Sci Inst (9):62–68. Translated in Spec Rep Fish, 327. U.S. Dep Inter, Fish and Wildlife Serv, Washington D.C., pp 69–76

Anderson LG, Drange H, Chierichi M, Fransson A, Johannessen T, Skjelvan I, Rey F (2000) Seasonal and annual variability in the upper Greenland Sea based on measurements and a box model. Tellus 52:1013–1020

Blindheim JV, Østerhus S (2005) The Nordic Seas, main oceanographic features, in The Nordic Seas. An integrated perspective. AGU Geophys Monogr 158:11–39

Blindheim J, Borovkov V, Hansen B, Malmberg SA, Turrell WR, Østerhus S (2000) Upper layer cooling and freshening in the Norwegian Sea in relation to atmospheric forcing. Deep-Sea Res I 47:655–680

Deser C, Magnusdottir G, Saravanan R, Philips A (2004) The effects of the North Atlantic SST and sea-ice anomalies on the winter circulation in CCM3. Part II: Direct and indirect components of the response. Journal of Climate 17:877–889

Dietrich G, Kalle K, Krauss W, Siedler G (1980) General oceanography, 2nd edn. Wiley, New York, p 626

Drange H, Dokken T, Furevik T, Gerdes R, Berger W (2005) The Nordic Seas: an integrated perspective, AGU Monograph vol 158. Washington, p 374

Druet C, Kowalik Z (1970) Dynamika Morza. Wydawnictwo Morskie, Gdańsk, p 428

European Environment Agency (2008) Impacts of Europe's changing climate. indicator based assessment, EEA–JRC–WHO Rep. No 4/2008

Flatau MK, Talley L, Niiler PP (2003) The North Atlantic Oscillation, surface current velocities, and SST changes in the Subpolar North Atlantic. J Clim 16:2355–2369

Furevik T, Nilsen J (2005) Large-scale atmospheric circulation variability and its impacts on the nordic seas ocean climate—a review. In: Drange H et al (eds) The Nordic Seas: an integrated perspective, Geophysical Monograph Series vol 158, AGU

Furevik T, Bentsen M, Drange H, Johannessen JA, Korablev A (2002) Temporal and spatial variability of the sea surface salinity in the Nordic Seas. J Geophys Res 107C12:8009. doi:10.1029/2001JC001118

Furevik T, Mauritzen C, Ingvaldsen R (2007) The flow of Atlantic Water to the Nordic Seas and Arctic Ocean. In: Arctic alpine ecosystems and people in a changing environment. Springer, Berlin. doi:10.1007/978-3-540-48514-8_8

Hansen B, Østerhus S, Hatún H, Kristiansen K, Larsen MH (2003) The Iceland-Faroe inflow of Atlantic water to the Nordic Seas. Prog Oceanogr 59:443–474

Hansen B, Østerhus S, Quadfasel D, Turrell W (2004) Already the day after tomorrow? Science 305:953–954

Hansen B, Østerhus S, Turrel W, Jonsson S, Valdimersson H, Hatún H, Olsen S (2008) The inflow of atlantic water, heat and salt to the Nordic Seas across the greenland-scotland ridge. Arctic-Subarctic Ocean Fluxes, Springer, Berlin, pp 15–43 (Science)

Hansen B, Østerhus S (2000) North Atlantic-Norwegian sea exchanges. Prog Oceanogr 45:109–208

Helland-Hansen B, Nansen F (1909) The Norwegian Sea, its physical oceanography based upon the Norwegian Research 1900–1904. Report Norwegian Marine Invest 2:390

Hurrell J (1995) Decadal trends in the North Atlantic Oscillation: regional temperatures and precipitation. Science 269(5224):676–679

Ingvaldsen R, Asplin L, Loeng H (2004) The seasonal cycle in the Atlantic transport to the Barents Sea during the years 1997–2001. Cont Shelf Res 24:1015–1032

Isachsen PE, LaCasce JH, Mauritzen C, Haakkinen S (2003) Wind-driven variability of the large-scale recirculating flow in the Nordic Seas and Arctic Ocean. J Phys Oceanogr 33:2534–2550

Jonsson S, Briem J (2003) Flow of Atlantic Water west of Iceland and onto the north Icelandic shelf. ICES Marine Sci Symp 219:326–328

Karcher MJ, Gerdes R, Kauker F, Köberle C (2003) Arctic warming: evolution and spreading of the 1990s warm event in the Nordic seas and the Arctic Ocean. J Geophys Res 108(C2):3034 (C001265)

Kalnay E, Kanamitsu M, Kistler R, Collins W, Deaven D, Gandin L (1996) The NCEP/NCAR 40-year reanalysis project. Bull Amer Meteor Soc 77:437–470

Manley TO (1995) Branching of Atlantic Water within the Greenland-Spitsbergen passage: an estimate of recirculation. J Geophys Res 100(C10):20627–20634

Massel S (2007) Ocean waves breaking and marine aerosol fluxes. Springer, Berlin, p 323

Mork KA, Blindheim J (2000) Variation in the Atlantic Inflow to the Nordic Seas, 1955–1996. Deep-Sea Res 47(6):1035–1057. doi:10.1016/S0967-06379900091-6

National Oceanic and Atmospheric Administration (2006) 2-minute gridded Global Relief Data ETOPO2v2, Nat. Geophys. DataCenter, U.S. Dept. Comm., Boulder, http://www.ngdc.noaa.gov/

Orvik KA, Niiler P (2002) Major pathways of Atlantic water in the northern North Atlantic and the Nordic Seas toward the Arctic. Geophys Res Lett 29(19):1896. doi:10.1029/2002GL015002 (L015002)

Orvik KA, Skagseth Ø (2003) The impact of the wind stress curl in the North Atlantic on the Atlantic inflow to the Norwegian Sea toward the Arctic. Geophys Res Lett 30(17):1884. doi:10.1029/2003GL017932

Orvik KA, Skagseth Ø (2005) Heat flux variations in the eastern Norwegian Atlantic Current toward the Arctic from moored instruments, 1995–2005. Geophys Res Lett 32:L14610. doi:10.1029/2005GL023487

Piechura J, Walczowski W (1995) The Arctic Front: structure and dynamics. Oceanologia 371:47–73

Poulain PM, Warn-Varnas A, Niiler PP (1996) Near-surface circulation of the Nordic seas as measured by Lagrangian drifters. J Geophys Res 101(C8):18237–18258

Rahmstorf S, Ganopolski A (1999) Long-term global warming scenarios computed with an efficient couplet climatic model. Climatic Change 43:353–367

Rudels B, Fredrich HJ, Quatfasel D (1999) The Arctic circumpolar boundary current. Deep-Sea Res II 46:1023–1062

Rudels B, Fahrbach E, Meincke J, Budeus G, Eriksson P (2002) The East Greenland Current and its contribution to the Denmark Strait overflow. ICES J Mar Sci 59:1133–1154

Schauer U, Fahrbach E, Osterhus S, Rohardt G (2004) Arctic warming through the Fram Strait: Oceanic heat transport from 3 years of measurements. J Geophys Res 109:C06026. doi:10.1029/2003JC001823

Schauer U, Loeng H, Rudels B, Ozhigin V, Dieck W (2002) Atlantic Water flow through the Barents and Kara Sea. Deep-Sea Research I 49:2281–2296

Schauer U, Beszczynska–Moeller A, Walczowski W, Fahrbach E, Piechura J, Hansen E (2008) Variation of measured heat flow through the Fram Strait between 1997 and 2006. Arctic-subarctic Ocean fluxes. Springer, NY, pp 15–43

Simonsen K, Haugan P (1996) Heat budgets of the Arctic mediterranean and sea surface heat flux parameterizations for the Nordic seas. J Geophys Res 101(C3):6553–6576. doi:10.1029/95JC03305

Skagseth Ø, Orvik KA, Furevik T (2004) Coherent variability of the Norwegian Atlantic Slope Current derived from TOPEX/ERS altimeter data. Geophys Res Lett 31:L14304. doi:10.1029/2004GL020057

Stewart RH (2006) Introduction to physical oceanography. Department of Oceanography, Texas A and M University

Steele M, Morley R, Ermold W (2001) PHC: A global ocean hydrography with a high quality Arctic Ocean. J Clim 14:2079–2087

Swift JH (1986) The Arctic waters. In: Hurdle BG (ed) The Nordic seas. Springer, Heidelberg, pp 129–153

Tomczak M, Godfrey JS (1994) Regional oceanography: an introduction. Pergamon, Oxford, p 400

Turrell WR, Hansen B, Haghes S, Østerhus S (2003) Hydrographic variability during the decade of the 1990s in the Northeast Atlantic and southern Norwegian Sea. ICES Mar Sci Symp 219:111–120

U.S. Department of Commerce, National Oceanic and Atmospheric Administration, National Geophysical Data Center (2006) 2-minute Gridded Global Relief Data ETOPO2v2

Walczowski W (1997) Transfrontalna wymiana masy i ciepła w rejonie Frontu Arktycznego, PhD Thesis, IOPAN, p 123

Walczowski W, Piechura J, Osinski R, Wieczorek P (2005) The West Spitsbergen Current volume and heat transport from synoptic observations in summer. Deep-Sea Res I 52:1374–1931

Walczowski W, Piechura J (2006) New evidence of warming propagating toward the Arctic Ocean. Geophys Res Lett 33:L12601. doi:10.1029/2006GL025872

Walczowski W, Piechura J (2007) Pathways of the Greenland Sea warming. Geophys Res Lett 34:L10608. doi:10.1029/2007GL029974

Wegner G (1973) Geostrophische Oberflächenströmung im nordlichen Nordatlantischen Ozean im Internationalen Geophysikalische Jahr 1957/58, Ber. Dt. Wiss. Komm. Meereforsch., 22, 411–426

Chapter 3
Methodology of Measurements and Data Analysis

3.1 Participation of the IOPAS in Research Programs and Measurement Campaigns

The Institute of Oceanology PAS has been conducting scientific research on the European Arctic Seas since 1987. Each summer, the Institute's research vessel 'Oceania' sails out for an expedition. The cruises are held under an original research program of the Institute of Oceanology—AREX. After years of experience, an optimum stations' grid seems to have been developed. The cruise includes measurements of physical and chemical properties of sea water and collection of hydrographic and biological samples. The present work discusses the research and includes some physical data gathered mainly during the AREX cruises in years 2000–2007. The author has participated in and led many of the expeditions as a chief scientist.

In the 1987–1993 period, the activities in the Arctic were conducted under a program called the Greenland Sea Project (GSP). The project was developed and coordinated by the AOSB and ICES. The GSP aimed at observing and modelling processes occurring in atmosphere, hydrosphere, cryosphere and biosphere, required for recognition of the climate-forming role of the Nordic Seas. The Institute of Oceanology took part in the project thanks to the initiative of the director of the Institute at the time—Professor Czesław Druet. The Institute conducted research of the Atlantic Water inflow through the Faroe-Shetland Channel (Schlichtholz and Jankowski 1993) and the region between the North Cape and Spitsbergen (Druet and Jankowski 1991, 1992).

When the GSP program terminated, the research activities were continued between Norway and Spitsbergen, and the research area was extended farther north. In addition to the large-scale quasi-synoptic surveys, initiated by the head of the Physical Oceanography Department—Professor Jan Piechura, a program of process-oriented measurements was commenced. In 1993 the Arctic Front research program was developed to study the area where the Atlantic-origin water and the Arctic water divide. A study of the frontal processes began in 1987 when RV "Oceania" crossed the Arctic Front for the first time; more detailed observations

W. Walczowski, *Atlantic Water in the Nordic Seas*,
GeoPlanet: Earth and Planetary Sciences, DOI: 10.1007/978-3-319-01279-7_3,

were performed in the period from 1993 to 1996 (Piechura and Walczowski 1995, 1996; Walczowski 1997).

In 1997 the Institute acceded VEINS—the international research program funded under the European Union's MAST-III programme. The immediate objective of the VEINS was to conduct measurements and modelling of the variability of exchanges between the Arctic Ocean and the Atlantic Ocean. A long-term objective was to develop and implement a system of measurements which would enable recognizing the role of the Arctic Ocean and the Nordic Seas in decadal climate changes. The VEINS program ended in 2000.

In 2000 IOPAS began hydrographical observations of the entire Atlantic Domain of the Greenland Sea (Fig. 3.1). In 2002 the ASOF program commenced. The Institute of Oceanology participated in the project as an equal partner, although Poland was not yet a member of the European Union. Considering our previous experience as well as the specific characteristic of our ship (RV "Oceania" is not suitable for ice navigation) we began our work under the ASOF-N part which mainly concerned the inflow and distribution of warm Atlantic Water in the Nordic Seas and its transport to the Arctic Ocean through the Fram Strait.

The measurements from the previous years, particularly those repeated at the same stations and transects proved to be very valuable. During the field campaigns we maintained the idea of covering as large area of the Atlantic Domain as possible with a grid of stations, and tightening the grid on the way up north. This

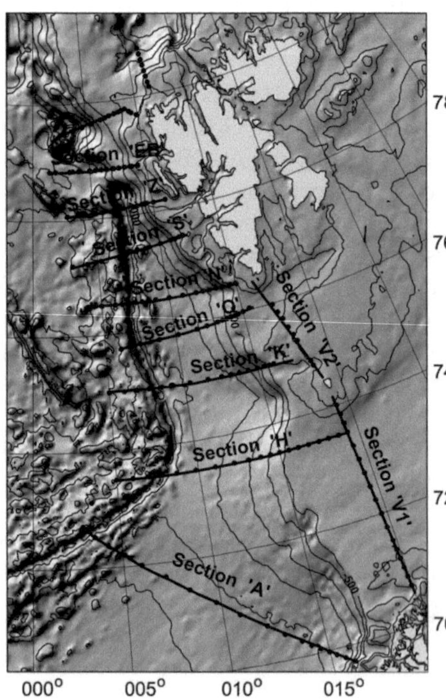

Fig. 3.1 Positions of the AREX 2004 measurement stations. The names of the sections most frequently measured are given

would provide a possibility of studying the inflow of the Atlantic Water into the Greenland Sea and the Barents Sea, modification of properties and transport of the AW on its way north, and, finally, the complex pattern of the West Spitsbergen Current.

The ASOF programme finalized in 2005. Thanks to the joint effort of the participants, it was possible to develop a project and obtain funds for continuation of the research. In 2006 the DAMOCLES program began. The Institute of Oceanology is one of the participants of the program.

3.2 Measuring Instruments

In time, as the main objectives and emphases of research have evolved, the study areas, research tools and fieldwork methods have changed. Until today, however, a Conductivity, Temperature, Depth (CTD) probe has remained the basic tool for continuous profiling of water column. Using the gathered results—conductivity of sea water, temperature and pressure—it is possible to calculate the basic properties of water masses. In the practice of oceanography, water temperature is given in degrees Celsius (°C), salinity calculated from electronically measured conductivity in practical salinity units (psu) and pressure in decibars (dbar).

Until 1993, CTD measurements had been conducted with use of a Canadian CTD probe—Guildline 8710, working at a frequency of 8 Hz, in a range of depths between 0 and 1,000 m.

In 1994, Seabird 9/11 system was purchased, which operates at a frequency of 24 Hz in a range of depths between 0 and 3,400 m. The equipment is a state-of-the-art pumped CTD system, most frequently used for oceanographic worldwide. Temperature measurement accuracy is 0.001 °C and conductivity is measured with an accuracy of 0.0003 S/m, which provides an average accuracy of salinity measurement of 0.003. The accuracy of piezoelectric pressure sensor is 0.015 % of the full depth range. Temperature and electrical conductivity sensors were sent for refurbishment and calibration to the manufacturer (USA) each year. During cruises, water samples were collected with use of a bathymetric rosette for calibration of conductivity in the laboratory. In 2005, a Guildline salinometer was installed aboard RV "Oceania" Oceania to measure samples during the cruise. In the same year electronic reversing thermometers were purchased for complementary temperature measurements. A double pair of sensors measuring temperature and conductivity has been used since 2006.

Each sampling was performed down to the bottom, the CTD probe was lowered with a velocity of approximately 1 m/s. At a frequency of sampling of 24 Hz, measurements were performed with an average interval of 4 cm. The measured data were transmitted online through a conductive sea cable, and recorded in a binary form on a hard disk.

The Guildline CTD probe was adapted for scanning the water column during the ship's movement, which allowed to perform high-resolution horizontal sections. A lot of sections were taken with use of the towed CTD system at the Arctic Front and in the shelf area of the Barents Sea and Spitsbergen. Since 2004, Seabird 49 has been used as a towed CTD scanfish.

An Acoustic Doppler Current Profiler (ADCP) manufactured by RDI was purchased in 1998. The device is installed in the bottom of the ship's hull. It operates at a frequency of 150 kHz and enables to measure sea currents during the ship's movement, in a layer of 350 m depth. While performing measurements in deep water, GPS data are used for determining the ship's movement, whereas in shallow water, the bottom reference is used. Additionally, Ashtech Attitude-sensing system (ADU) has been used since 2003 to improve determination of the ship's heading, pitch and roll motion (Osinski et al. 2003).

Work Horse Sentinel ADCP 300 kHz, manufactured by RDI, was purchased in 2003. It is an ADCP equipped with internal memory and custom software for vertical profiling (Lowered ADCP—LADCP) of sea currents. The measurement is performed during a standard CTD profiling. It is the first device which allows for continuous measurement of sea currents from the surface to the bottom, from aboard. The measurements are performed while lowering and raising the profiler. CTD logs also contain GPS navigation data concerning the ship's movement while profiling.

Throughout the years of work, the major IOPAS research infrastructure—RV Oceania—has also been improved. In 1994 the first state-of-the art GPS navigation device was installed to receive and log the ship's position data while performing towed CTD measurements. The major difficulty of deep water measurements was eliminated in 1997: old hydrographic winches, manually-controlled from the deck were replaced by modern electronically-controlled cable winches with levelwinds. In 1998 the bottom of the ship's hull was reconstructed and an ADCP was installed. In 2003 ADU was installed, which greatly improved the quality of data gathered with ADCP. In 2001 a hydraulic crane of 1000 kg capacity was installed on the board, and lateral winches were replaced by A-frames. Concurrently, laboratories were rearranged—mainly the data collection laboratory—to adapt them to an increasing amount of equipment and changes in the structure of computers.

3.3 The Study Area

The result of the IOPAS activity in the Arctic is a large database. Until 2007 it contains more than 3300 CTD stations, continuous ADCP measurements since 2003, and 1000 LADCP profiles. Its advantage is continuity—the measurements have been performed at the same stations for years, always during the same season. After years valuable time series have been collected, allowing to commence the study of the variability of the properties, transport and circulation of water masses.

The additional advantage of the collected data is the fact that a quasi-synoptic image of the Atlantic Domain has been taken since 2000. That is the only in the world such a complete and consistent dataset enabling to study the structure and dynamics of water masses in the Atlantic Domain of the Nordic Seas. However, besides its merits, the synopticity of observations has also some disadvantages. The collected time series cover only the summer season. Thus, when studying temporal variability on the basis of those data, it must not be called ‘interannual variability’ but ‘summer-to-summer variability’ at most. While processing the collected data, the importance and value of multiyear time series may be appreciated, for example those provided by moorings. On the other hand, the ‘synoptic image’ taken each year, allows to study the processes occurring downstream of the moorings at the Svinoy Section and upstream of those installed in the Fram Strait, and then compare the obtained results. Thanks to participation in the international programs, we can cooperate with foreign partners and have access to various data they have collected.

The unchanged grid of measurements repeated since 2000, as well as systematic observations conducted in the same season, allowed to collect a consistent data set—a time series of hydrographic images of the AREX study area. The data are supplemented each year and their cognitive value is increasing. Due to the time period of measurements—usually from 20 June to 20 July—the data are not strictly synoptic. On the other hand, time scales of processes which occur in the ocean vary a lot—from minutes to hundreds of years. After an adequate processing, the measured data might be used to analyse the change in hydrographic conditions of a given sea region as quasi-synoptic data. However, it must be emphasized again that by using the data, one can only refer to interannual variability for a given season—summer—in that particular case. Extrapolating of the results to the whole year is risky, although in the case of integrated values or values averaged with great volume, it can be done, with adequate limitations.

Measurement strategy was based on the performance of cross-sections perpendicular to presumed direction of current (Fig. 3.1). Transects arranged along parallels intersected the most important bottom structures which limit the Atlantic Domain—the shelf and slope of the Barents Sea on the east and the chain of underwater ridges on the west. Two meridional transects closed the entrance to the Barents Sea. In individual years the extent of transects, particularly towards the west and north, varied due to changing ice situation, however, Fig. 3.1 pictures their general arrangement well. The horizontal resolution of the profiles was different. In the areas of particular current intensity, mesoscale activity or hydrographic fronts, the distances between the profiles were reduced, even to 10 km, whereas in the areas of weaker variability, the stations were 30–35 km apart. Also the distance between the transects varies. The southern part of the study area is not as well covered by transects as the northern part. In that region, conclusions as to processes, including in particular dynamic ones, must be drawn with great caution.

3.4 Characteristics and Methods of Processing the Measurement Data

Approximately 1,500 vertical CTD profiles, performed by IOPAS in the period 2000–2007, were used in the above-presented results. Additionally, to close the study areas from the south, data from the Fugloya-Byornoya Section 2005, the Gimsoy Section 2006 and the Gimsoy Section 2007, collected by the Institute of Marine Research in Bergen, Norway (52 stations) were used.

Routines of standard processing of data from the CTD profiling were introduced. The data from measurement were processed in stages. The raw data were first converted from the binary form into ASCII files, then spikes were filtered out, temperature and conductivity sensors' reaction times were compensated, low-pass filter was applied, pressure slowdowns or reversals were eliminated, vertical averaging applied and derived parameters were calculated. The properties directly measured with a CTD probe include: pressure, temperature and conductivity of sea water. Using dedicated software, the values of potential temperature, salinity, potential density, specific volume anomalies, Brunt-Vaisala frequency, as well as other parameters were calculated. All the variables were standard-averaged onto 1 or 5 m intervals.

The data from the acoustic Doppler current profiler (ADCP) were processed with use of procedures described by Osiński (2000). While processing, the values of current direction and velocity were standard-averaged in 5-minute ensembles, which results in the resolution of 0.75 km at the vessel's speed of 5 knots.

The data obtained with use of the LADCP (using the GPS data logged during the CTD profiling) were processed with software developed by Visbeck and Krahmann (Thurhnerr 2005). When the processing was complete, homogenous (barotropic) components of tidal currents were filtered. K1, O1, M2 and S2 tidal current constituents were subtracted. U and V components of tidal currents were calculated with use of amplitudes and phases modelled by Kowalik and Proshutinsky (Kowalik 1994, Kowalik and Proshutinsky 1995). LADCP current sections were obtained from vertical profiles gridded along transects with use of the same methods as the CTD sections and baroclinic currents.

Geostrophic currents were calculated with use of a dynamic height method, with reference to the no motion layer (NML) of 1,000 dbar or to the bottom. In order to improve the baroclinic calculations, velocity reference obtained from the ADCP data is applied for some transects.

The processing—data extraction and preparation, projection of geographical coordinates into a cartesian grid, and so on, were performed in the Matlab package. The operations of interpolation and visualization of the results were performed in the Surfer package. In case of the horizontal distributions (maps) the data were interpolated on a rectangular 12 × 12 km grid. Kriging procedures were applied as the interpolation objective analysis method (Emery and Thomson 2001). The interpolated data were smoothed with a low-pass averaging filter. In the case of vertical profiles, anisotropic factor was also used. The study uses both smoothed and unsmoothed fields. Equal-area map projection was applied to transform the measurements positions on the Earth's surface into locations on a plane. Flow velocities, volume and heat transport as well as average parameters of water masses were calculated from the interpolated data.

Distributions (maps) of the following physical parameters on the selected isobaric levels were developed on the basis of obtained measurement data:

- Potential temperature (T);
- Salinity (S);
- Specific volume anomaly (SVA);
- Geopotential Anomaly (GPA);
- Velocities of baroclinic currents (Vel);
- Kinetic energy of baroclinic currents (KE).

Furthermore, maps of the following mean or vertically integrated properties of the Atlantic Water layer were analysed:

- Potential temperature (T);
- Salinity (S);
- AW layer thickness (Th);
- Heat content of the AW layer (Q).

First, the mean value (temperature, salinity, density) or integrated value (layer thickness, heat content) was calculated for individual measurement profiles vertically, and then the points values were interpolated spatially. Such prepared data were used to calculate mean values and standard deviations of each parameter for summers 2000–2007. Furthermore, anomalies of the properties for each summer season were calculated. The area of study common for all the years, hereinafter referred as "the AREX study area" has an area of 313,654 km^2. The investigations extended beyond the boundaries of the AREX area in individual years and horizontal distributions for such years are presented for larger areas, however, all the mean values, standard deviations and anomalies presented in tables and diagrams are calculated for the common AREX study area, as shown in Fig. 4.1, in order to maintain the possibility of comparison. Also the properties of water in smaller areas or on hydrographic sections were analysed. In individual years, the lengths of the transects differed, for instance, due to various ice conditions. When calculating mean values and comparing the results, the transects, similarly to horizontal distributions, were limited to the common part covered by measurements in all the years.

To analyse the correlation between various physical properties of the AW, the linear regression method was applied. Pearson's linear correlation coefficient r was calculated from the following formula (Bendat and Piersol 1976):

$$r = \frac{\sum_{i=1}^{N} (x_i - \bar{x})(y_i - \bar{y})}{\sqrt{\sum_{i=1}^{N} (x_i - \bar{x})^2 \sum_{i=1}^{N} (y_i - \bar{y})^2}} \tag{3.1}$$

where: N—number of measurements, x_i, y_i—values of variables x, y at the point i (for i = 1 … N), $\bar{x}$, $\bar{y}$—mean values of the variables x, y.

Furthermore, p-values were calculated to test the hypothesis of no correlation. In that case the p value means the probability of random obtaining a given correlation value with no actual correlation between the studied values. Due to short time series, the p-value was calculated with use of t-statistics for N - 2 degrees of freedom. If the p-value is lower than 0.05, the correlation is significant (significance level 0.05), and in the text the value r is bolded. If the p-value is lower than 0.01, the correlation is defined as very significant (significance level 0.01) and the value r is bolded and underlined. When calculating correlation coefficients, detrended time series were used. This guarantees elimination of artificial overestimation of the coefficient. As a rule, such calculated correlation coefficient is lower than a coefficient calculated for raw data. However, sometimes the opposite occurs—the correlation coefficient calculated for detrended series is higher. Such a situation may take place when trends of both variables are opposite. This results in problems with interpretation of the results. A decision was made to present such data with reservations. Raw data with trends were used for preparing correlation and linear regression diagrams. It was deemed to generate more natural and intuitive results, and the correlation coefficients for detrended data are provided for comparison in suitable tables or in the text.

All the statistic calculations, including in particular the results of correlation analysis, should be approached with most caution and far-fetched conclusions should not be drawn on their basis. First of all, time series are short. Only further observations may verify some ideas and suggestions. Secondly, one should always bear in mind that a correlation between two values does not always have to implicate a cause-and-effect relationship. It is a frequent case that both values are modified by an external (not always known) process.

The Empirical Ortogonal Functions (EOF) are essentially useful when investigating time series of geophysical data spatial distribution. The advantage of the EOF is the possibility of recording N spatial synoptic measurements of the value Z (x, y, t) as independent spatial components (patterns) EOF (x, y) and time series of amplitudes PC(t):

$$Z(x, y, t) = \sum_{i=1}^{N} PC(t) * EOF(x, y) \tag{3.2}$$

The terminology in that case is not homogenous and quite troublesome. In the present work, horizontal distribution (maps) of amplitudes or correlations shall be defined as EOF (with respective number), whereas 'series'—time series of the EOF shall be determined as PC (Principal Components).

The work presents EOF for the series of distribution of the AW layer mean temperature and salinity. The EOF were calculated from gridded and smoothed fields. As regards the data processed herein, short time series of the data can be considered as a disadvantage, which is also the case when calculating correlations.

3.5 Measurement Errors and Data Processing Errors

Errors and inaccuracies are unavoidable and occur at all stages of work—starting from gathering of data and ending with their presentation. Measurement errors include mostly inaccuracies resulting from accuracy and resolution of sensors measuring water conductivity, temperature and pressure. In that case they were reduced to the minimum—the CTD SeaBird 911+ sensors were calibrated by the manufacturer before each cruise. Calibration curves were entered into the computer in the profiler data acquisition software. Greater errors may occur when sending data from the profiler to computer via the conductive sea cable and sliprings of the hydrographic winch. Although the signal is in a digital form, it does not eliminate errors which mostly occur as the so-called spikes. Most of the errors were filtered in the profiler data acquisition software and the rest is generally easy to remove by data filtration and averaging while processing. Nevertheless, failures of the data acquisition system sometimes caused (as in 2006) the necessity of editing data, manual removal of spikes and labour-consuming processing of each profile separately.

However, one may claim with certainty that point CTD casts provide high-quality data and they fulfil all the international standards. The accuracy of temperature and salinity measurement are at least one order of magnitude higher than those required for climate studies.

Other errors result from the fact that the data are gathered within the period of one month. Thus, the presented distributions, mean values, etc. are not synoptic data, but only quasi-synoptic. In that case it is assumed that processes occurring in the ocean are slow enough ('the ocean memory') to let us treat the data as simultaneous. An additional advantage that allows for comparing the results is the fact that the measurements were always performed in the same season and in the similar sequence of stations.

Subsequent errors occur while interpolating the point data spatially. This inevitably leads to inaccuracies—both in the points of measurement—the gridded value is different from the measured one, and in the points where there are no measurements, and the value of the grid results only from interpolation. Mapping the actual hydrographic fields is certainly worse in the parts of the study area where fewer profiles are taken. Thus, it is much worse in the southern part of the study area, south of the 74°N parallel.

The present work deals mainly with large-scale processes occurring over a long period of time. Therefore, in the majority of cases, horizontal distributions were smoothed with low-pass filter. This leads to an increase of residuals for point measurements (Table 5.1), but, on the other hand, it allows to obtain smoother fields and eliminate mesoscale phenomena.

Next, mean salinity, temperature and other AW properties over the entire AREX study area, its northern and southern part, were calculated on the basis of smoothed gridded fields. Based on the average residuals (Table 3.1), conclusions may be drawn as to the magnitude of errors while estimating mean AW

Table 3.1 Residuals T_{dat}—T_{grd} for 177 temperature and salinity data of the AW layer in 2006. Data for a non-filtered grid and a grid filtered with low pass filter

Grid	Average matching errors	Standard deviation	Minimum	Maximum
Non-filtered temperature	0.0025	0.2063	−0.9306	0.6243
Filtered temperature	0.0401	0.5253	−1.2046	2.2463
Non-filtered salinity	−0.0002	0.0122	−0.0524	0.0463
Filtered salinity	0.0055	0.0321	−0.0847	0.0851

temperature and salinity for the entire study area. Such integrated data are less sensitive to potential errors which occur at each stage of data collection and processing, and they reflect climate processes best. Another serious problem is the question if and how the data collected during one season may translate onto a whole year. The purpose here is not to calculate, e.g. winter data on the basis of summer data, but to determine trends—for example, whether whole-year data reveal the same trend as the data collected in summer. Since there are no whole-year data and adequately long series, a decision was made to run an experiment on model data. For this purpose, the Polar Ice Prediction System (PIPS) (Maslowski et al. 2000)—high resolution ocean and ice model, developed and run in Monterey (California)—was used. Data representing the transect "N" of the AREX study area were extracted from averaged monthly results of the model. Two series of calculations were performed—for full annual data from the period between 1979 and 2005, and summer data (July) of the same period (Fig. 3.2).

The monthly data clearly reveal high seasonal variability. Temperature maximum occurs, similarly as in nature, in autumn months—October or November. Variability and high peaks of salinity are surprising. The diagram obtained from the summer data somehow modifies the temperature and salinity curves. This should be no surprise, as the sampling frequency is 12 times lower. Nevertheless, the general image of warming and cooling periods is maintained. In both cases, linear trends of both temperature and salinity are positive (Table 3.2). The trends obtained from the July data are slightly lower, which suggests that determination of trends from summer in situ collected data will not lead to over-estimation of the actual trends. This complies with the general description of climate change in the Arctic. It is frequently emphasized that changes of temperature (e.g. air temperature) in winter are much greater than in summer.

It must be emphasized here that results of the model have not been used for direct comparison with the measurement data, but only for the purpose of verifying if the measurements are representative.

Table 3.2 Values of linear trends for the transect 'N', data from the numerical model

Trend	Whole-year data	Data from July
Temperature (°C/10 years)	0.1694	0.1535
Salinity (1/10 years)	0.0082	0.0065

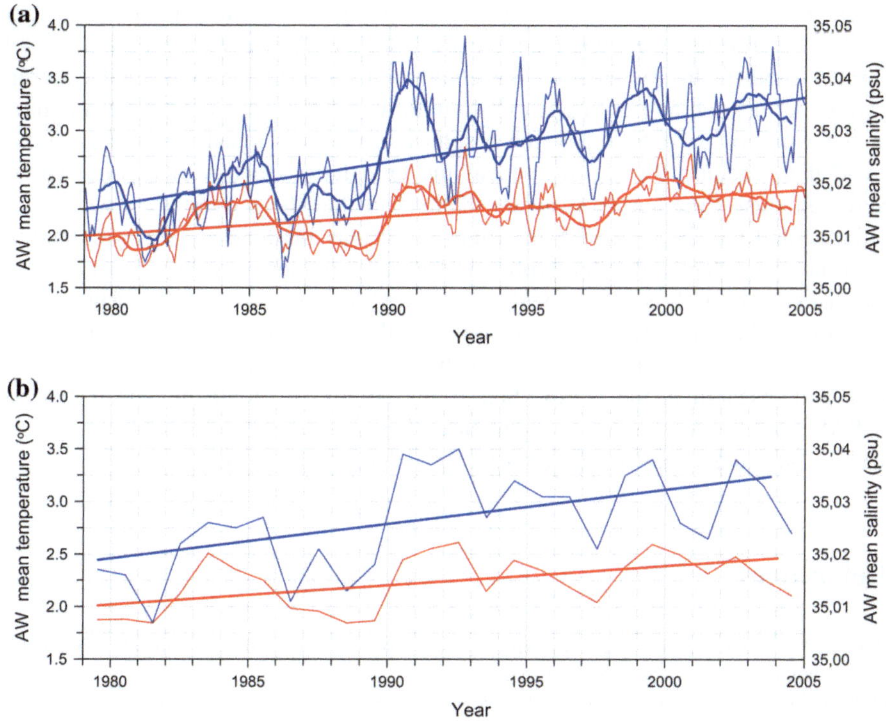

Fig. 3.2 Time series of salinity and temperature for transect 'N' obtained from the model. *Red line*—temperature, *blue line*—salinity. The *thick lines* show the moving averages (11 months and 11 years respectively) and the linear trends. *Upper figure*—all the monthly means, *lower figure*—means for July

3.6 Problems and Doubts

The primary problem of the present work is the shortage of data. It may sound paradoxical in a situation when thousands of vertical profiles and millions of point values of temperature and salinity were used. 25 data sets were acquired for each metre of depth. Additionally, ADCP and LADCP data were collected. Atmospheric data, satellite images of sea surface and ice cover are available. We have access to the high resolution numerical model of the Arctic Ocean. Yet, the deficiency of data can be still felt. On the other hand, the AREX study area is equal to the area of Poland. Furthermore, it is three-dimensional—the depth of the described Atlantic Water layer is several hundred metres. Once a year, the study area is covered by 200 measurement points in a non-uniform arrangement. Based on that, attempts are made to recognize and describe dynamic processes that take place in the region and draw conclusions as regards the occurring changes. The most serious deficiency is the lack of data from other seasons—the continuity of data is only conventional, since the frequency of measurements is low.

On the other hand, a problem with excess of data occurred. While preparing the work, new results were obtained during a cruise aboard R/V Oceania, which slightly modify the picture and the elaborated ideas that had seemed coherent before. A question appeared what to do with them. They could be either included in the work or used to verify the results and the developed ideas. The author decided to include the results, repeat the calculations and verify some views. Simultaneously, the author is aware of the fact that the presented views are not ultimate and will be verified many times. That may be done mostly by extending the time series. Thus, it is the problem of years. Another possibility is to enhance the data base with measurements performed by other institutes. The author abandoned that, he refers to them only if additional data are indispensable. The reason is that one of the intentions of the work was to present the IOPAS' data base and research in the Nordic Seas. This obviously should not mean that the IOPAS study in that region is independent and self-contained wealth. It shall be seen as an integral part of the entire knowledge about the Arctic.

References

Bendat JS, Piersol AG (1976) Metody analizy i pomiaru sygnałów losowych. Państwowe Wydawnictwo Naukowe, Warszawa

Druet C, Jankowski A (1991) Flow across south and east boundaries of the Norwegian Sea. Oceanologia 30:37–46

Druet C, Jankowski A (1992) Some results of three—year investigations on the interannual variability of the Norwegian—Barents confluence zone. Polish Polar Res 13(1):3–17

Emery WJ, Thomson RE (2001) Data analysis methods in physical oceanography. Elsevier Science B.V, Amsterdam

Kowalik Z (1994) Modeling of topographically amplified diurnal tides in the Nordic Seas. J Phys Oceanogr 24:1717–1731

Kowalik Z, Proshutinsky YA (1995) Topographic enhancement of tidal motion in the western Barents Sea. J Geophys Res 100(C2):2613–2637

Maslowski W, Newton B, Schlosser P, Semtner A, Martinson D (2000) Modeling recent climate variability in the Arctic Ocean. Geophys Res Lett 27(22):3743–3746

Osiński R (2000) The misalignment angle in vessel-mounted ADCP. Oceanologia 42(3):385–394

Osinski R, Wieczorek P, Beszczynska-Moller A, Goszczko I (2003) ADCP-referenced geostrophic velocity and transport in the West Spitsbergen Current. Oceanologia 45(3):425–435

Piechura J, Walczowski W (1995) The Arctic Front: structure and dynamics. Oceanologia 371:47–73

Piechura J, Walczowski W (1996) Interannual variability in the hydrophysical fields of the Norwegian-Barents Seas confluence zone. Oceanologia 38:81–98

Schlichtholz P, Jankowski A (1993) Hydrological regime and water volume transport in the Faeroe-Shetland Channel in summer of 1988 and 1989. Oceanol Acta 16(1):11–22

Thurnherr AM (2005) LADCP processing how-to, for version IX of the LDEO processing software, based on Visbeck's V8b

Walczowski W (1997) *Transfrontalna wymiana masy i ciepła w rejonie Frontu Arktycznego*. PhD thesis, IOPAN, p 123

Chapter 4
Properties of the Atlantic Water in 2000–2007

4.1 Mean Properties of the Atlantic Water

The present study features definition of the Atlantic Water based on the criteria of minimum temperature and minimum salinity. The AW is considered as water with temperature $T > 0$ °C and salinity $S > 34.92$. The selection of parameters itself might be disputable. The AW is conventionally considered as watermass with temperature $T > 3$ °C and salinity $S > 35$. However, having entered the Nordic Seas, the AW releases heat to the atmosphere and cools down, mixes with surrounding waters and becomes fresher, yet still maintaining its signature. That is the reason why the parameters T and S are often decreased in that case (Hopkins 1991). Carmack and Aagaard set the lower limit of the AW temperature as 0 °C already in 1973. The author uses the above mentioned definition in the majority of papers, since these values concide with the layer of strongest vertical gradients separating the AW from the water masses below.

The figures below show mean values and standard deviations (SD) of salinity, temperature, and heat content in the AW layer in the AREX study area in the investigated period. They are weighted means, with thickness of the AW layer, presented in a separate distribution, used as the weight for each year.

The AW physical properties change meridionally along the direction of the flow as well as zonally, across the AW domain. The AW entering from the south is characterized by higher salinity and temperature; the water is transformed during its northward advection. As a result of mixing with the surrounding waters, the AW on the peripheries of the domain is less saline and colder. Heat exchanges between ocean and atmosphere take place over the entire area, which results in decrease of the AW temperature along its northward flow. Salinity does not vary significantly on seasonal time scale therefore is less susceptible to changes resulting from exchanges with the atmosphere than temperature. Distribution of mean salinity of the AW layer (Fig. 4.1) is therefore strongly related to the AW inflow into the Greenland Sea. A wide inflow of water with mean salinity above 35.08 enters between the northern Norwegian coasts and the system of underwater ridges. At the eastern side, closer to Norway, salinity is lower due to the activity of

W. Walczowski, *Atlantic Water in the Nordic Seas*,
GeoPlanet: Earth and Planetary Sciences, DOI: 10.1007/978-3-319-01279-7_4,

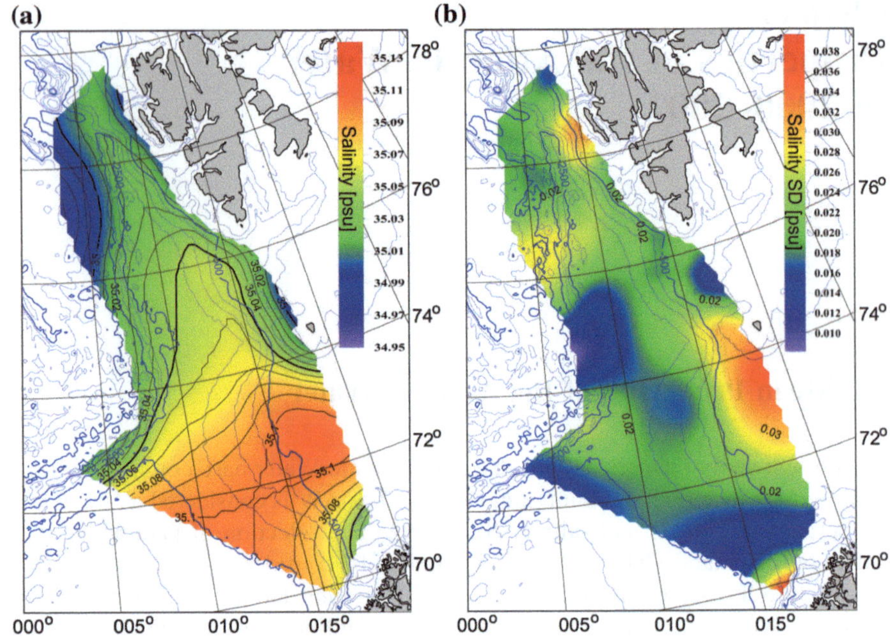

Fig. 4.1 Mean salinity of the AW layer in July 2000–2007 (**a**) and its standard deviation (**b**)

the Norwegian Coastal Current, whereas at the western side lower salinity occurs in the frontal area of the Arctic Front. The high salinity area is situated towards the north-east, along the direction of the AW inflow into the Barents Sea. North of the high salinity gradients area, above the 74°N parallel, the AW flow is oriented meridionally. The Atlantic Domain (AD) narrows down as the bottom topography changes. From the east the flow is limited by the shelves of the Barents Sea and the west Spitsbergen, and from the west by the Mohn Ridge and the Knipovich Ridge. As the AW flows northward, its mean salinity decreases due to mixing with surrounding less saline Polar and Arctic waters.

The strongest variations of salinity (high value of standard deviation) occurs in the area of the Barents Sea shelf, south of the Bear Island. The variability is related to the Polar Water inflow from the Barents Sea. Another area of increased variability is north of the 76°N parallel, in the western side of the study area, and north of 78°N, near the Spitsbergen coast. Increased variability in the former area is connected with the AW recirculation variability in that region, while in the latter region it results probably from the convergence of two WSC branches and local impacts of the West Spitsbergen fjords.

Mean temperature of the AW changes from 6.2 °C in the south-east of the study area to 2 °C in the north-west (Fig. 4.2). In the southern and central area the horizontal temperature gradient is directed to the north-west, while in the Spitsbergen area the gradient of mean temperature is directed to the north. The warm

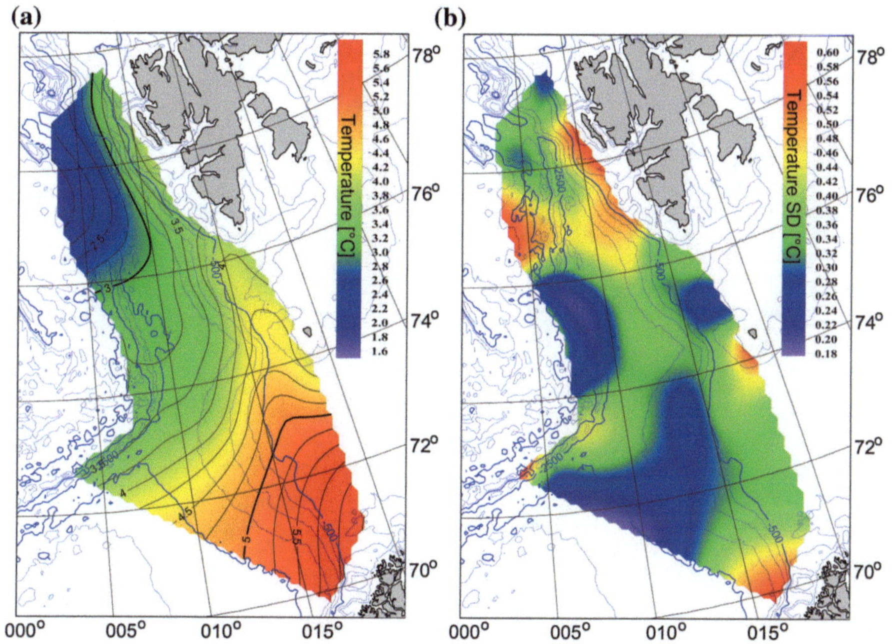

Fig. 4.2 Mean temperature (°C) of the AW layer in July 2000–2007 (**a**) and its standard deviation (**b**)

AW is clearly visible in the northward flow along the continental slope, representing the core of the West Spitsbergen Current, while coldest AW is found in the area of recirculation and bifurcation of the Arctic front.

It is significant that there are no temperature gradients towards the north-west, along the flow, in the WSC core area, above the Spitsbergen shelf. That might mean that the transport is fast enough to prevent the water from cooling down in summer. The areas of increased standard deviation of temperature overlap analogical areas for salinity.

In the central part of the Atlantic Domain, the AW layer thickness reaches above 500 m on average (Fig. 4.3). The AW layer is locally thicker even by 100 m. Two branches of the flow can be recognized in the AD shape—a current along the Barents Sea shelf (15°E meridian), with a branch flowing into the Barents Sea, and a flow along the underwater ridges. The both WSC branches converge at 78°N latitude and then the AW flow diverge. The West Spitsbergen Current divides here into a number of branches (Manley 1995). Figure 4.3 clearly shows the inflow of the Svalbard Branch into the Arctic Ocean, along the continental shelf break and the westward recirculation.

The strongest variability of AW thickness occurs in the frontal zones: near the north Norway, in the area where the AW contacts the Norwegian Coastal Current and in the Arctic Front area. It is particularly in the second case that a tendency to

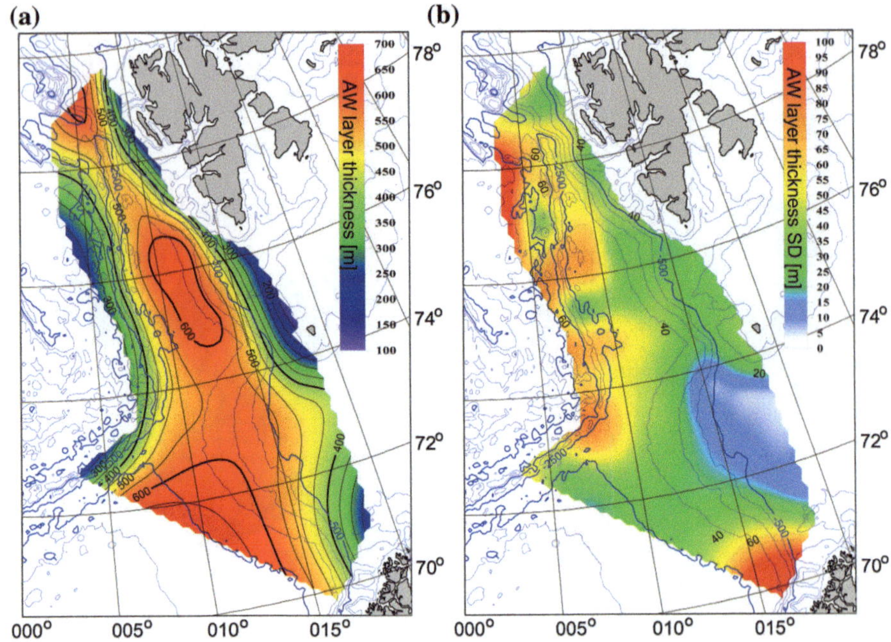

Fig. 4.3 Mean AW layer thickness (m) in July 2000–2007 (**a**) and its standard deviation (**b**)

change of the front position and intensive mesoscale activity are reflected by the AW layer thickness variability.

Considering the processes of transport and heat exchange, heat content distributions in the AW layer (Fig. 4.4) is the most interesting parameter since it combines information about the AW layer thickness and temperature. In the southern part, 1 m^2 of the AW column contains up to 12 gigajoules (1 GJ $= 10^9$ J) of heat, whereas in the northern part the value is reduced by half. The width of the Atlantic Domain also decreases strongly, thus the amount of heat accumulated in the AW and transported northward is significantly reduced. Between the latitudes of 74°N and 76°N there is an area of nearly constant heat content (8 GJ/m^2); farther south there is a zone of strong meridional gradients. Similarly to the AW thickness, the 78°N parallel is an area of some kind of discontinuity, significant narrowing of the AW flux.

4.2 Changes in the Properties of the Atlantic Water

The Atlantic Water salinity and temperature reveal significant changes in the short period of 8 years, between June 2000 and July 2007. Between 2000 and 2003 both values varied strongly and a sudden increase occurred in 2004. The increase lasted

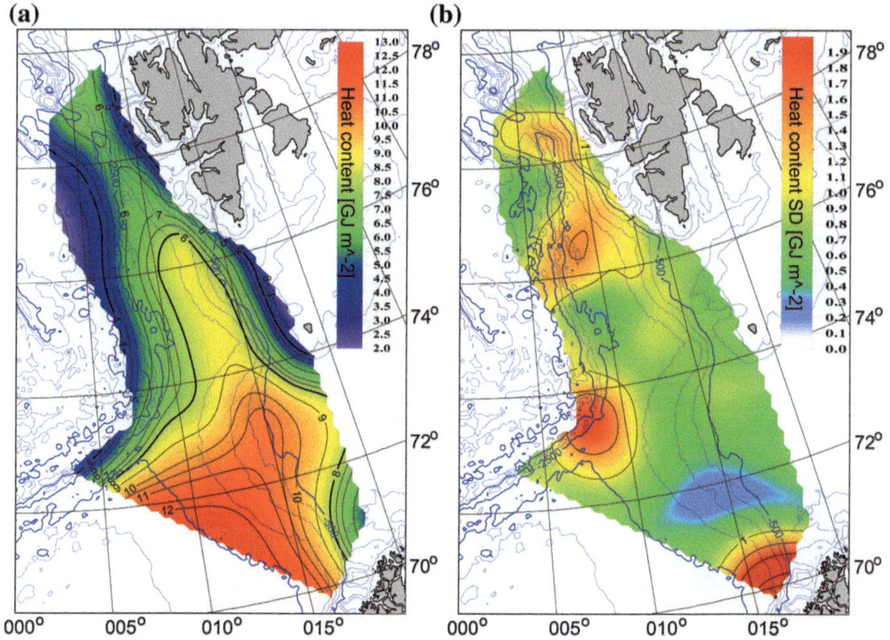

Fig. 4.4 Mean AW heat content (GJ m^{-2}) in July 2000–2007 (**a**) and its standard deviation (**b**)

until 2006 and in 2007 the both values decreased suddenly. This is indicated by both the data from single oceanographic stations, mean properties of the AW from individual sections and averaged values for the entire water mass.

While describing the AW variability, horizontal distributions of averaged (Figs. 4.5, 4.6) or integrated (Figs. 4.7, 4.8) properties in the AW column are shown for each year, as well as time series of averaged or integrated parameters for the entire study area. In the case of the temperature and salinity series, these are weighted means (the AW layer thickness in a given location is used as the weight), and the heat content is an integral from the entire AREX study area. In addition to the figures, Table 4.1 also contains mean values of the discussed AW properties for the common AREX study area. Table 4.2 contains correlation coefficients between the individual AW properties. The data trend may have serious influence and may artificially increase the correlation coefficient. Therefore, calculations were made for detrended time series of mean values from the study area.

Mean salinity of the AW layer is the first presented time series (Fig. 4.5). In the beginning the salinity increase was insignificant, then in the period 2004–2005 the salinity rose suddenly. In summer 2006 there was a slight decrease of mean AW salinity (by 0.002) and in 2007 the decrease was already significant and mean AW salinity fell to the level before 2004. A series of horizontal distributions shows the dynamics of changes, the initial transport of the more saline AW towards the north and west, withdrawal in 2003 and then expansion towards the Fram Strait.

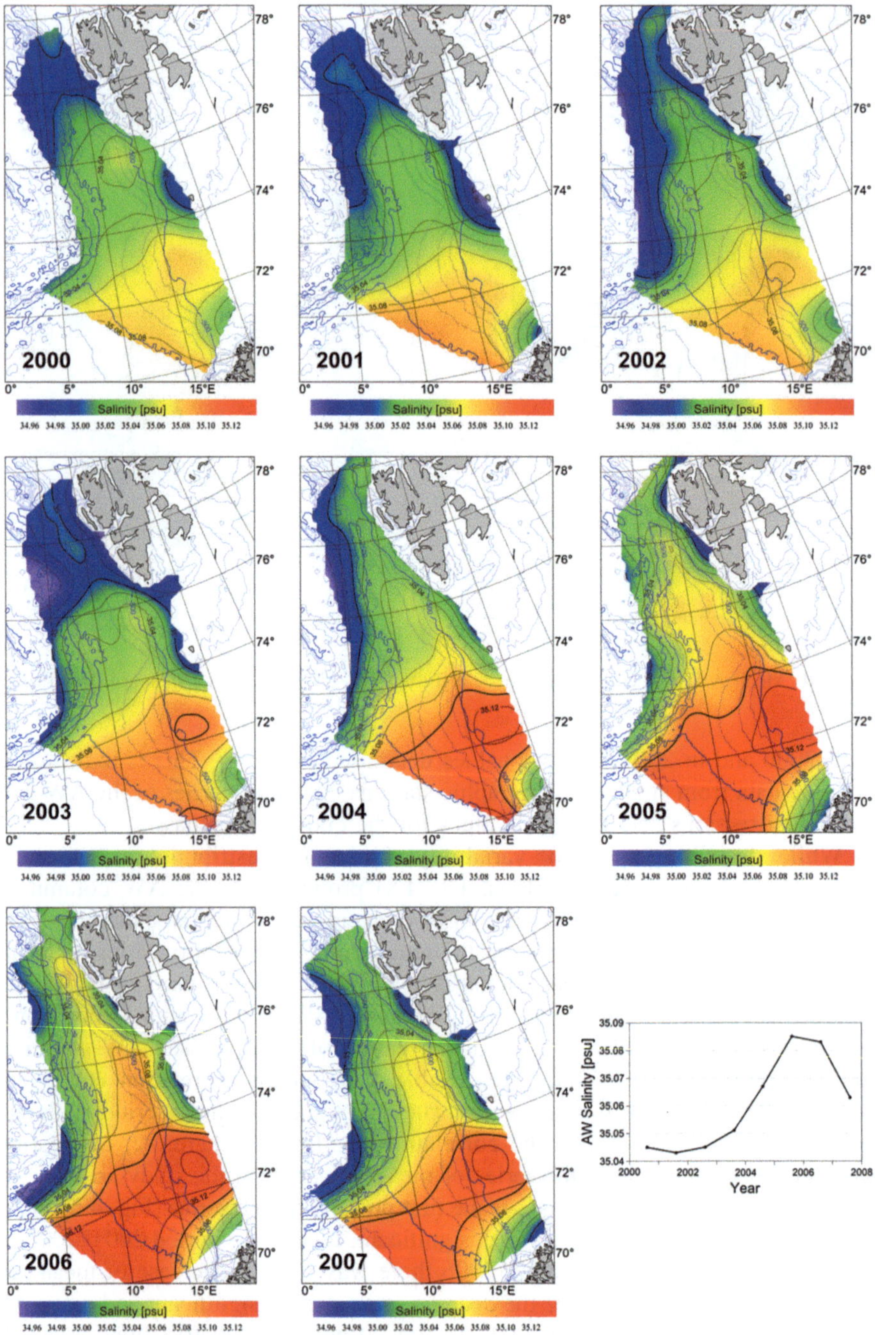

Fig. 4.5 Horizontal distributions of vertically averaged AW salinity in subsequent summer seasons, and time series of average weighted salinity of the AW layer for the AREX study area. 35.00 and 35.05 isolines are *bold*

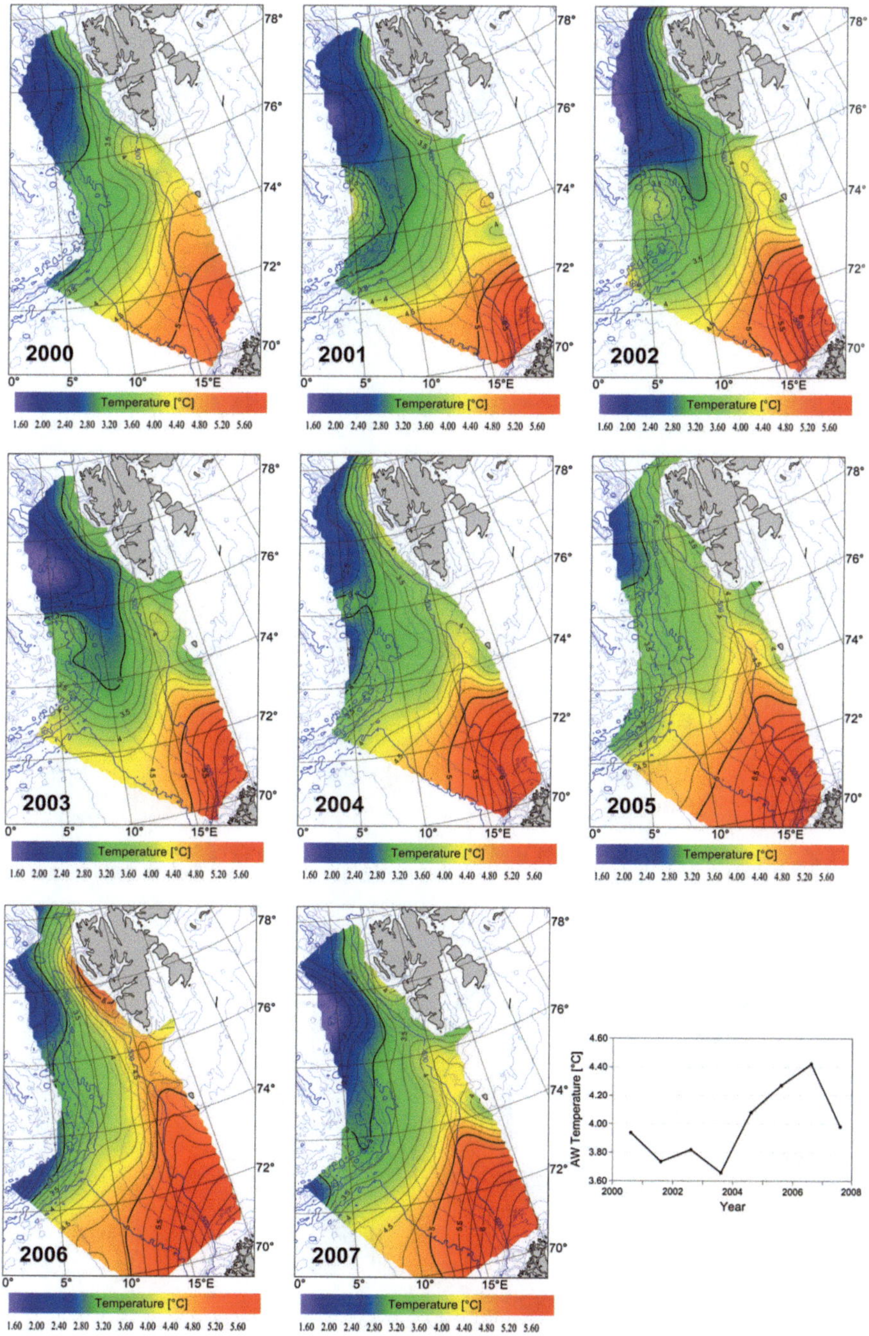

Fig. 4.6 Horizontal distributions of vertically averaged AW temperature (°C) in subsequent summer seasons, and time series of average weighted temperature of the AW layer for the AREX study area. 3 and 5 °C isolines are *bold*

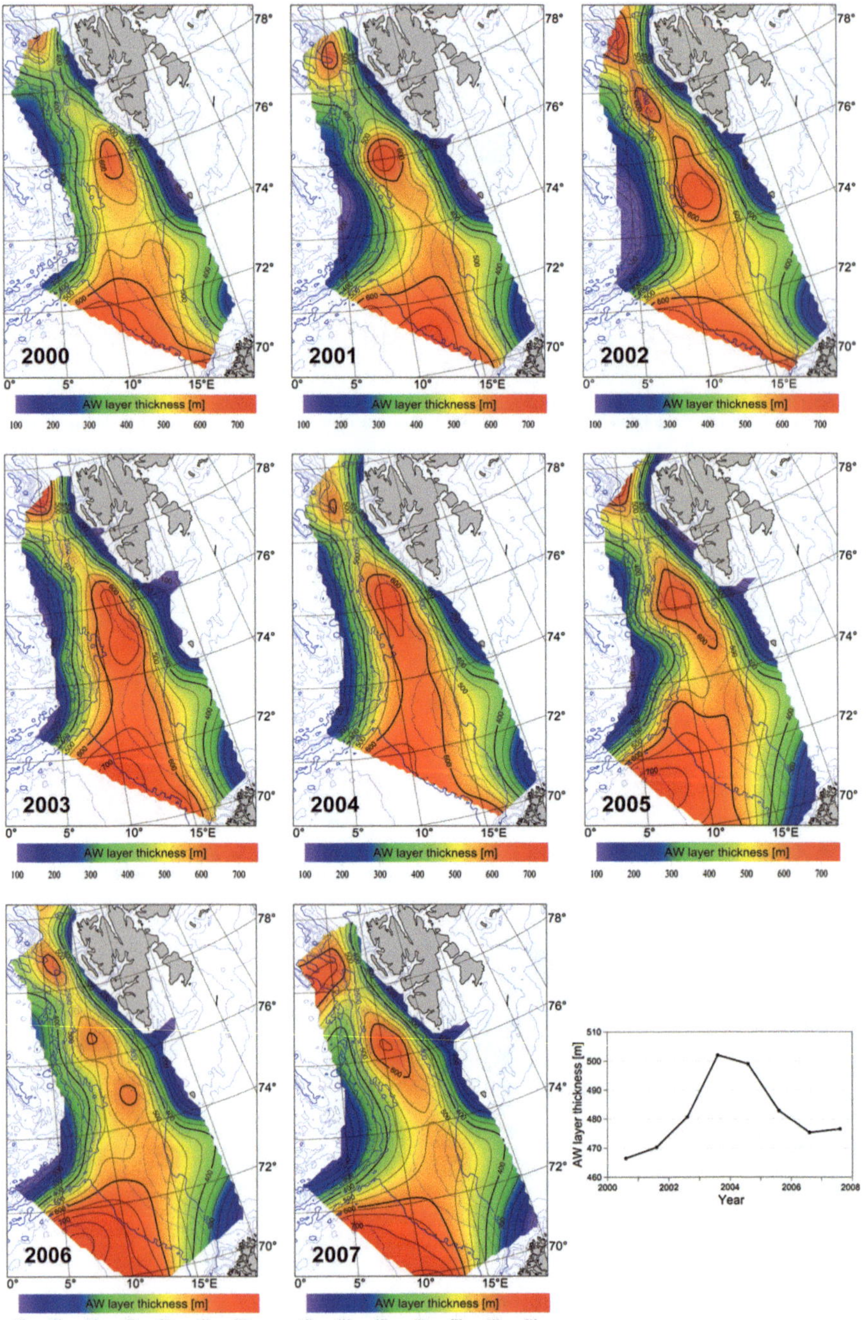

Fig. 4.7 Horizontal distributions of the AW layer thickness (m) in subsequent summer seasons and time series of mean AW layer thickness for the AREX study area. 400 and 600 m isolines are *bold*

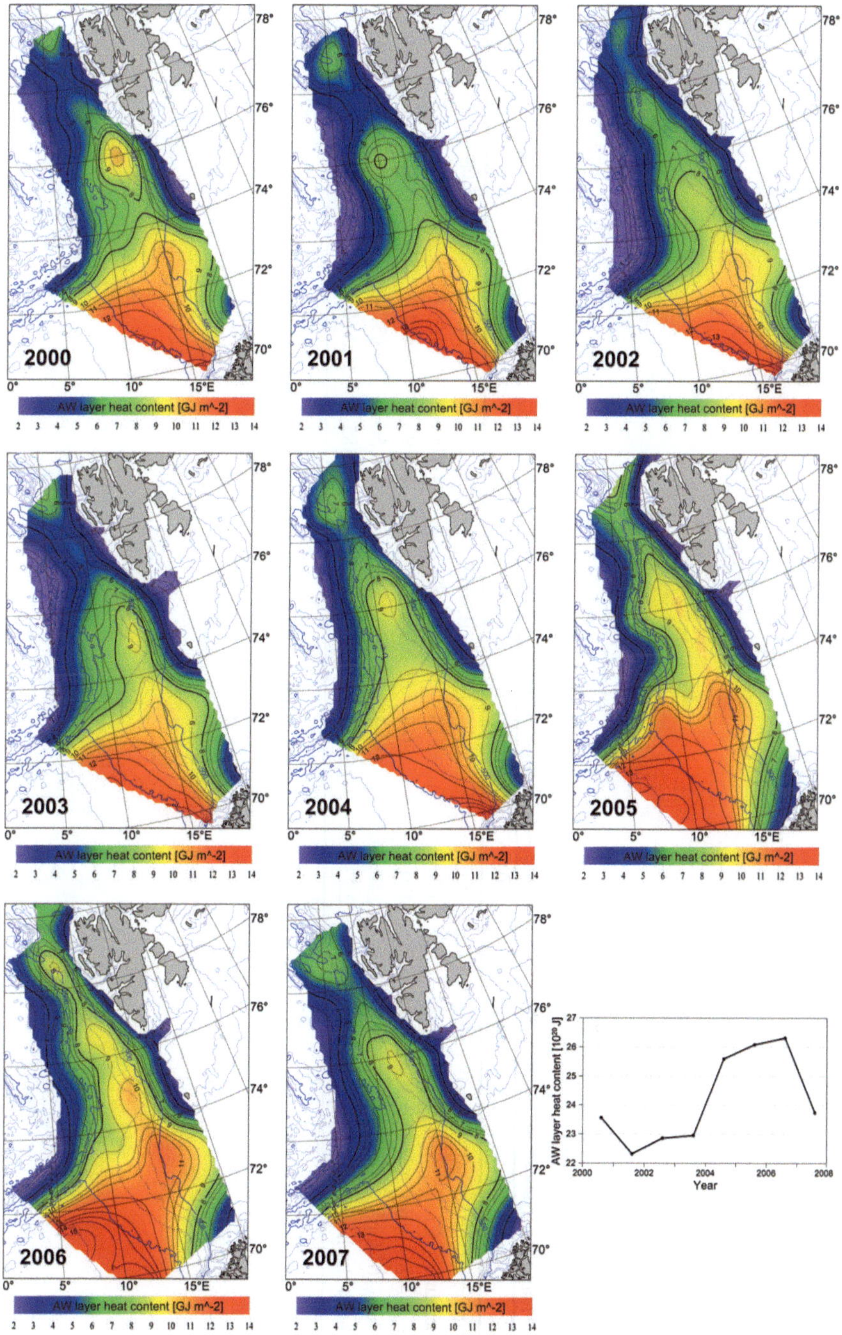

Fig. 4.8 Horizontal distributions of heat content (GJ m^{-2}) in the AW layer in subsequent summer seasons and time series of total heat content (10^{20} J) in the AW layer for the AREX study area. 4 and 8 GJ m^{-2} isolines are *bold*

Table 4.1 Mean properties of the Atlantic Water layer in the AREX study area

Parameter	2000	2001	2002	2003	2004	2005	2006	2007
Mean salinity	35.045	35.043	35.045	35.051	35.067	35.085	35.083	35.063
Mean temperature (°C)	3.937	3.734	3.818	3.657	4.078	4.271	4.421	3.980
Mean layer thickness (m)	466.5	470.2	480.7	501.9	498.9	482.8	475.3	476.5
Layer volume (10^3 km^3)	146.31	147.48	150.77	157.43	156.50	151.45	149.07	149.45
Heat content in the layer (10^{21} J)	2.3553	2.2311	2.2851	2.2942	2.5585	2.6073	2.6305	2.3728

Table 4.2 Correlation coefficient *r* between individual mean properties of the AW for the entire AREX study area. Time series have been detrended

Parameter	AW salinity	AW layer thickness	AW heat content
AW temperature	**0.87**	−0.38	**0.91**
AW salinity		−0.03	**0.93**
AW layer thickness			−0.02

Table 4.3 The Correlation coefficient *r* of time series of the AW mean properties in the AREX study area, the southern part

Parameter	AW salinity	AW layer thickness	AW heat content
AW temperature	0.53	−0.28	0.63
AW salinity		0.58	**0.92**
AW layer thickness			0.56

Regression occurred in 2007. The process is well visible as the change in the location of 35.00 and 35.05 isohalines. In 2003 the 35.05 isohaline reached the latitude of 74°30′N, and 35.00 the latitude of 76°45'N. In 2004 the both isolines shifted northwards by more than 2° of latitude (ca. 220 km), in 2006 the 35.00 isohaline shifted beyond the study area, and the 35.05 isohaline reached as far as the Fram Strait. In summer 2007 the meridional range of AW with mean salinity of 35.05 returned to the values from 2004. The processes will addressed in more detail in Chap. 6.

The Atlantic Water temperature is positively correlated with salinity (Table 4.3, the period of salinity increase is also accompanied by the increase of temperature. The temporal evolution of mean temperature (Fig. 4.6) is, however, slightly different than the temporal variability of mean salinity; they are most similar in the period of sudden increase (2004-2005). In 2006 the mean AW salinity already decreased a little, whereas temperature continued to rise; in 2007 the both values decreased suddenly. Horizontal distributions of mean temperature show the inflow of the warmest AW from the south-eastern part of the study area and its transport to the Barents Sea and towards the Fram Strait. The isotherms of 3° and 5°C (shown in bold) clearly indicate the expansion of unusually warm Atlantic waters. A particularly high variability can be seen on the western side of the Atlantic Domain, in the region of the Arctic Front. In that case the changes are frequently connected with mesoscale activity in that region, meandering of the front, creation and separation of anticyclonic eddies into the Greenland Sea, as well as intrusions of the Arctic Water into the Atlantic Domain (Walczowski 1997). The coldest years—2001 and 2003—are a period of significant cooling from the north-west. In the period 2005–2006 the AD clearly expanded westward. Weighted mean of water temperature in the Atlantic layer within the AREX study area increased by 0.75 °C (from 3.66 to 4.42 °C) from the summer of 2003 to the summer of 2007. In 2007 the values returned to those recorded in 2004.

Another presented value is the AW layer thickness (Fig. 4.7). That parameter also evolves in time. The AW thickness reached maximum (502 m) in 2003, when its temperature and salinity were the lowest. Horizontal distributions, despite being smoothed with a low-pass filter, show areas of locally increased AW thickness, characteristic for mesoscale structures. The image showing the years of the largest AW volume—2003 and 2004—is the most "continuous". Before and after that period, mesoscale activity can be clearly seen, the shape of AW stream is irregular. One may hypothesize that until 2003 the AW accumulated in the studied area and since 2004 there was a period of increased northward transport, since the mean thickness (and volume) of the AW layer decreases. To level the difference of 5,050 km^3 of the AW volume between 2005 and 2004, it is enough to increase the mean annual flow through the Fram Strait into the AO by the value of 0.16 Sv. Although huge resources are involved in studying oceanic exchanges through the FS, the value of this order is difficult to measure. Furthermore, one should bear in mind that a change in the AW volume also results from transport through the eastern and the western boundary. Also the flow dynamics has its annual cycle, increase or decrease of transport may occur in short time, and the AW thickness may increase or reduce several times per year.

Having only summer data available, conclusions must be drawn with caution, however, the diagram of the AW layer thickness (Fig. 4.7), similarly to the temperature and salinity time series, suggests the occurrence of decadal quasi-periodicity. That would confirm the thesis presented by Dukhovskoy et al. (2004, 2006a, b) concerning the periodicity of exchange through the Fram Strait. The authors state that the exchange is a type of natural oscillations of the system of 10 years period.

Although the AW volume decreased after 2003, total heat content in the Atlantic Water within the AREX study areas rose (Fig. 4.8). The volume decrease was compensated by significant increase of temperature. Very high correlation coefficient between temperature and heat content in the AW layer also shows that temperature plays a more important role in heat accumulation than layer thickness.

The distributions of the AW layer thickness and heat content in the West Spitsbergen region draws attention. The fields narrow down near the 78°N parallel and widen northwards. That is the above-mentioned "discontinuity" of fields caused by the convergence of the eastern and western WSC branches, acceleration of the flow and then divergence of the currents and deceleration of flows.

The highest correlations, both at the significance level of 0.01, were obtained between the following pairs of properties: temperature-heat content and salinity-heat content. The salinity-temperature pair is correlated at the significance level of 0.05. It is symptomatic that there is no direct correlation between mean temperature or salinity and the AW layer thickness or volume. It seems that the AW layer thickness may be connected rather with the dynamics of flow than with the AW physical properties. In some years the flow through the Fram Strait decreases. The lower transport of the AW into the AO may result in accumulation and increase of the AW layer thickness in the studied area. That may be indicated by the results obtained after the time shift was applied. It occurred that $r = \mathbf{-0.85}$ for the pair:

salinity (2000–2006)—layer thickness (2001–2007). The same coefficient of correlation exists between the pairs: temperature (2000–2006)—layer thickness (2001–2007). That might mean that the flow of warmer, more saline AW will reduce the thickness and volume of the AW in the following year. According to the simplest working hypothesis, the warmer AW the larger the steric heights, the faster baroclinic flows and, consequently, the thinner the AW layer.

4.3 Variability Studied Using the Empirical Orthogonal Functions

Time series of horizontal distributions (maps) of the AW layer temperature and salinity in the studied sea area were used to calculate the Empirical Orthogonal Functions. Variability of those properties in time was separated into spatial maps of amplitudes (EOF) and time series of principal components (PC).

The first EOF mode describes 68.0 % of total variance of the mean AW layer temperature, the second and the third—13.1 and 9.3 % of the variance, respectively. Since the EOF1 describes the major part of variability, the PC1 time series is very much similar to the time series of the AW layer mean temperature (Fig. 4.9). The changes in temperature occur simultaneously in the entire area of study; the amplitude of changes is higher in the northern part. The amplification effect is visible here. The temperature maximum is in 2006. The second mode (EOF2) shows the variability of similar period as in the first one, however, it divides the study area into the eastern and the western part, where the changes occur out of phase. It reveals the strongest temperature increase in 2000 and 2005 in the western part of the northern study area. The third mode (EOF3) also divides the area of study into the eastern and the western part, however, it shows variability on longer period.

Altogether, the first three EOFs describe 85 % of the temperature field variability in the AREX study area. A contribution of each mode to the total variance is well represented by the maps showing the correlation of measurement time series with a respective PC time series (Fig. 4.10). The first mode correlates best with the measurements in the central part of the Atlantic Domain, the other two represent variability on its margins. The third mode is strongly correlated with the temperature over the Barents Sea shelf—in the BSO and in the Storfjord Trench (up to −0.7). That may be connected with the variability of the AW exchange with the Barents Sea.

In the case of salinity, the first three EOF modes describe as much as 93 % of the AW layer mean salinity variability: 79.2, 8.8 and 4.7 %, respectively. Therefore, PC1 (Fig. 4.11) is also similar to the diagram showing the variability of the layer salinity. EOF2 divides the study area into the southern part and the northern part above the 74°N parallel. EOF3 divides the study area into the eastern and western part. The analysis of the correlation (Fig. 4.12) between EOF and

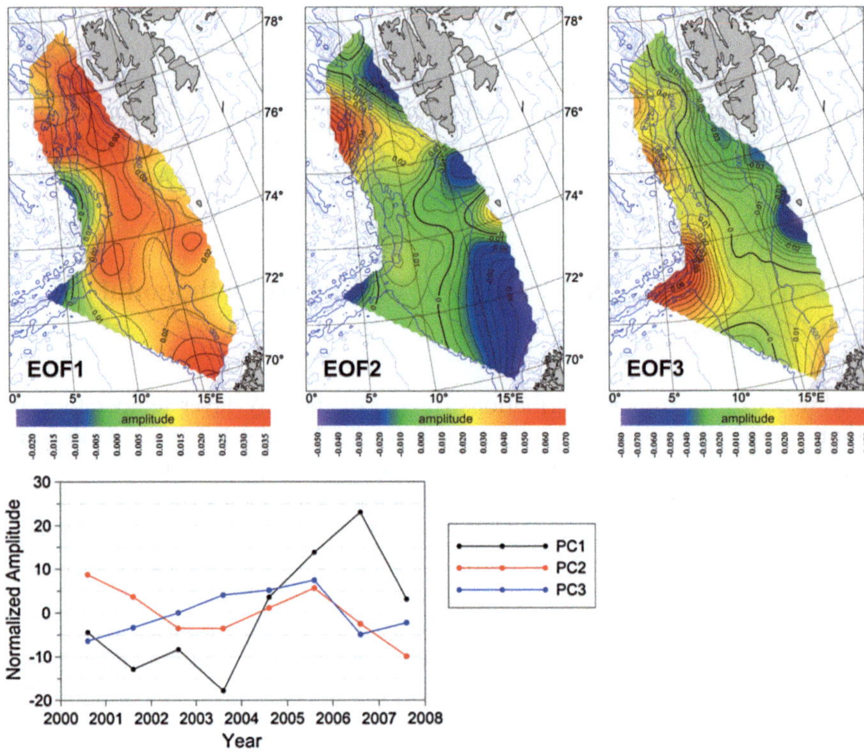

Fig. 4.9 The first three EOF modes and PC of the AW layer mean temperature

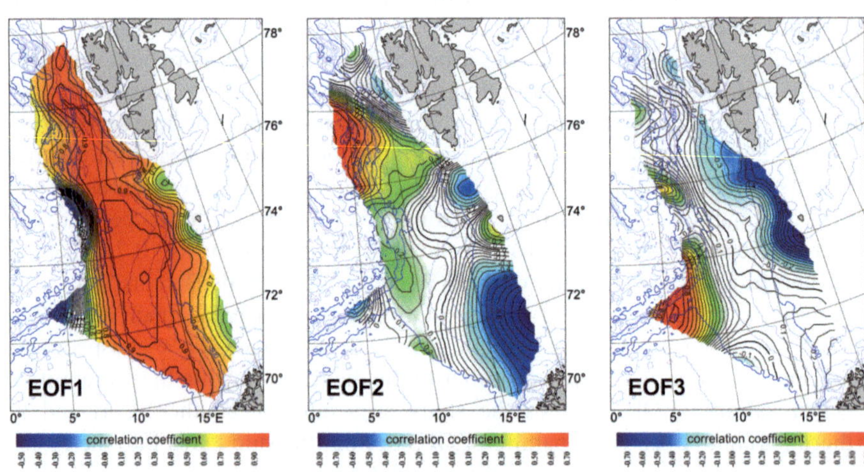

Fig. 4.10 The first three EOF modes of the AW mean temperature shown as spatial distributions of the correlation coefficients

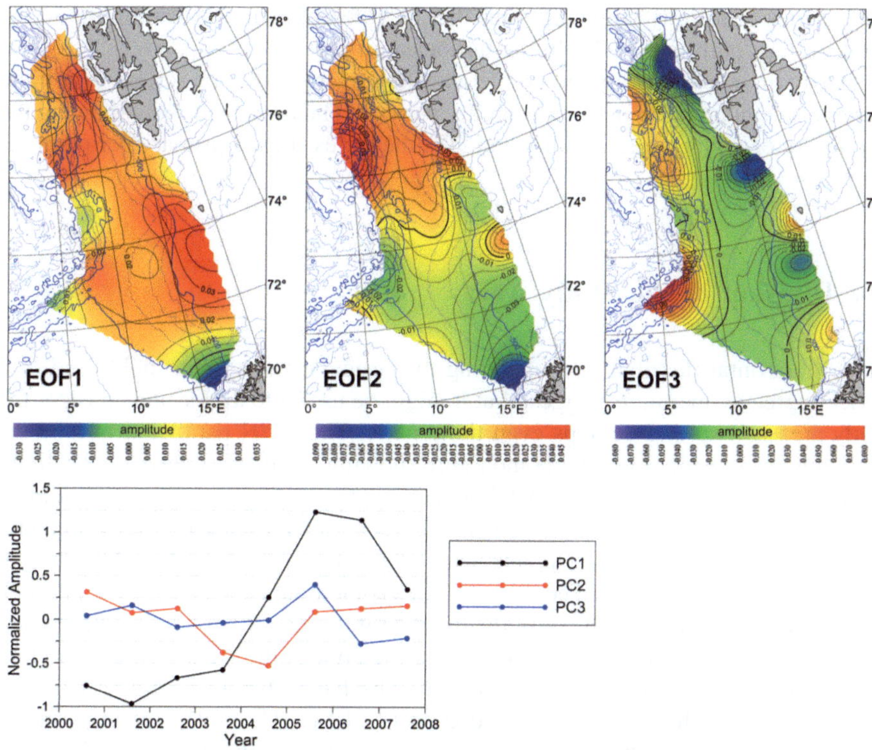

Fig. 4.11 The first three EOF modes and PC of the AW layer mean salinity

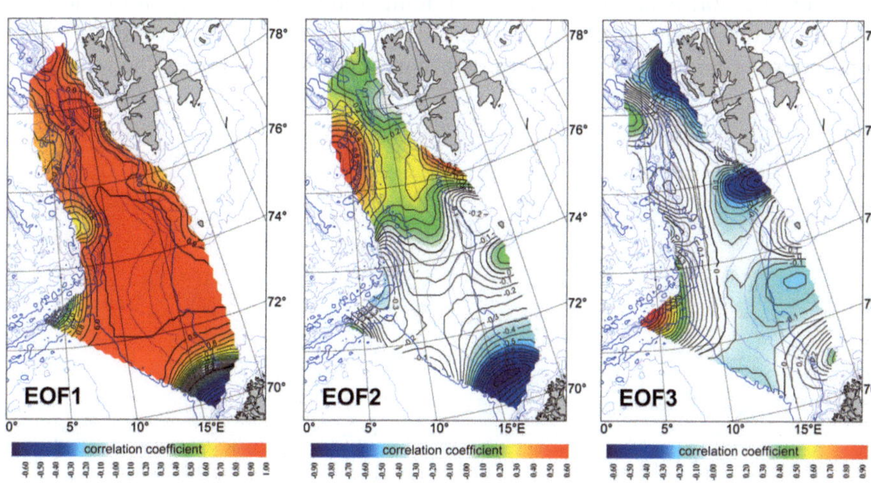

Fig. 4.12 The first three EOF modes of the AW mean salinity shown as spatial distribution of the coefficients of correlation

times series of salinity fields indicates that PC2 describes mostly variability in the northwestern part, and PC3—in the northeastern part.

Due to the above-mentioned problems with the length of time series, no definite statements can be made on the variability periods. The presented plots showing the variability of temperature, salinity and PC time series show a primary period of ca. 10 years.

4.4 Division of the Study Area

Some horizontal distributions, including in particular the salinity fields and to a lesser degree temperature, suggest that it might be useful to divide and consider the study area as two and perhaps even three parts. The 74°N parallel would be the main line of division. Looking at the figure showing the most conservative salinity field it is easy to see the distinct boundary between the zonal and meridional patterns of isohalines (Figs. 4.1, 4.5). The 74°N parallel divides the study area into the area where flows into the Barents Sea prevail (the southern part) and the area of flow into the Arctic Ocean. The second well visible boundary is the 78°N parallel. At the present stage of study, only the division into the northern and southern part by the 74°N parallel was taken into account. It divides the AREX study area into two equal parts of 156,827 km^2 area each. The basic AW layer properties were calculated for each part separately (Figs. 4.13, 4.14). The AW properties in the southern part are denoted with "S" index, and in the northern part—"N".

The values of all the four studied parameters (mean temperature, salinity, AW layer thickness, heat content in the AW layer) are larger in the southern area of study than in the northern one. This confirms that modification of water masses occurs during northward advection. Both temperature and salinity in the northern area of study change in phase with those in the southern area of study. Their amplitudes are larger in the northern part than in the southern one. The increase of

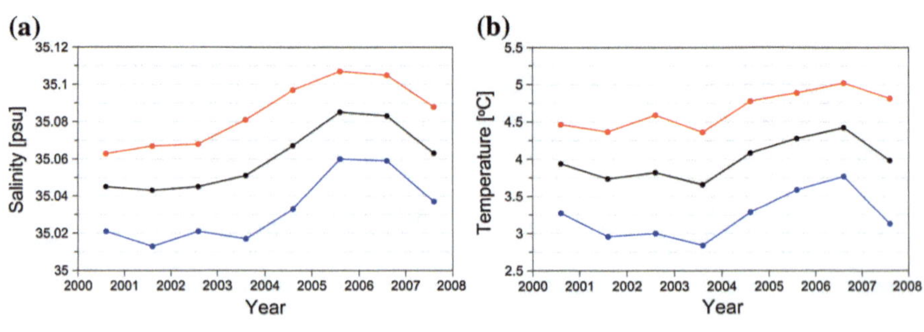

Fig. 4.13 Mean salinity (**a**) and temperature (**b**) of the Atlantic Water in the entire AREX study area (*black line*), the southern part (*red line*) and the northern part (*blue line*) in July 2000–2007

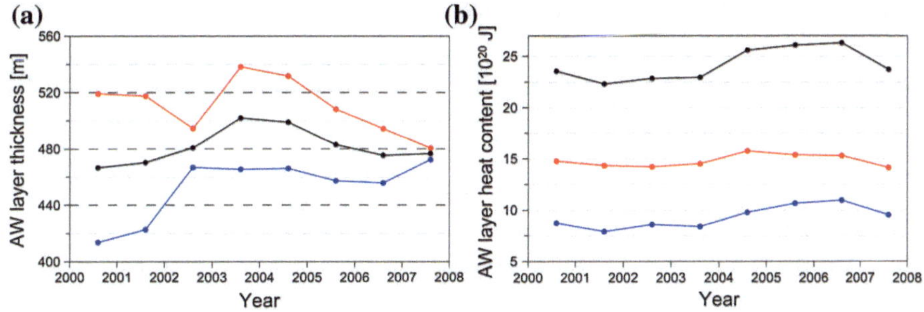

Fig. 4.14 The AW layer thickness (**a**) and heat content (10^{20} J) (**b**) in the entire AREX study area (*black line*), the southern part (*red line*) and the northern part (*blue line*) in July 2000–2007

salinity between 2004 and 2005 and a sudden decrease of temperature in 2007 is particularly pronounced in the northern part.

The change in the thickness (and volume) of the AW layer in the southern and northern part (Fig. 4.14a) is surprising. Changes occur out of phase, the AW layer thickening in the north coincides with reduction of thickness in the south. In 2007 the AW layer thickness in the northern part increased significantly and reached a level similar to the value in the southern part. Considering similar areas of the both parts, that means that the AW volume in the northern and southern part of the AREX study area also evened.

Furthermore, heat content in both areas (Fig. 4.14b) changes in phase with temperature. However, the layer thickness has a noticeable impact upon the heat content; in the northern part a significant decrease of the AW temperature in 2007 was compensated by the increase in the AW layer thickness, therefore, the heat content reduction was not so dramatic. Heat content remained above the mean value from the period 2000–2007. The decrease of temperature and salinity, with concurrent increase of the AW layer volume in the northern part in 2007, was presumably caused by intensive flow of the AW from the western branch in which the AW is less saline and colder.

The relationships between various properties of the AW layer, separately for the southern study area (Table 4.3), the northern study area (Table 4.4), and mixed correlations of properties from the southern and the northern part (Table 4.5) were checked. Time series have been detrended.

Statistically significant correlation coefficients are higher in the northern part (Table 4.4) than in the entire study area (Table 4.2) and in the southern part (Table 4.3). Temperature changes in the southern part are the most unusual, since they are poorly correlated with salinity and heat content there. On the other hand, salinity well correlates with heat content. Obviously, one should not presume that the change in salinity causes the change in heat content which is also a function of the AW volume.

The correlations rise significantly in the northern part. That may suggest that the Atlantic water mass in that region is more homogenous.

Table 4.4 The coefficient of correlation *r* of time series of the AW mean properties in the AREX study area, the northern part

Parameter	AW salinity	AW layer thickness	AW heat content
AW temperature	**0.94**	−0.57	**0.97**
AW salinity		−0.45	**0.95**
AW layer thickness			−0.37

Table 4.5 The coefficient of correlation *r* between time series of the AW mean properties from the southern (S) and the northern (N) part of the AREX study area

Parameter	AW-N temperature	AW-N salinity	AW-N layer thickness	AW-N heat content
AW-S temperature	**0.89**	**0.84**	−0.30	**0.90**
AW-S salinity	0.67	**0.73**	−0.13	**0.78**
AW-S layer thickness	−0.09	−0.07	0.20	0.05
AW-S heat content	0.70	0.66	−0.16	**0.80**

Temperature, salinity and heat content in the northern study area correlate with temperature in the southern study area without a time lag, all the correlations are at the level of significance 0.01 (Table 4.5). Salinity and heat content in the northern study area correlate with salinity in the southern study area at the level of significance of 0.05. The other correlations remain at the level of significance higher than 0.05. However, other interesting relationships occur between individual values with 1 year time lag:

SS (2000-2006)—TN (2001-2007) $r = \mathbf{0.79}$

SS (2000-2006)—ThN (2001-2007) $r = \mathbf{-0.71}$

QS (2000-2006)—ThN (2001-2007) $r = \mathbf{-0.77}$

ThN (2000-2006)—ThS (2001-2007) $r = \mathbf{0.91}$

The first three correlations seem logic: some phenomena occur in sequence from the south towards the north with one-year time-lag. However, the correlation coefficient between the AW layer thickness in the south and in the north is an exception to the rule. It demonstrates that the change of the AW layer thickness in the north leads its change in the south by one year. That may confirm the previous hypothesis that the layer thickness (AW volume) depends on the dynamics in the northern part of the study area and fluxes in the Fram Strait.

Based on temporal variability of the AW properties in the southern and the northern part, and having calculated the correlations in various configurations, the following conclusions may be drawn:

1. The temperature changes in the northern and southern part are in phase (Fig. 4.15); the coefficient of correlation is high (**0.89**). A similar relationship

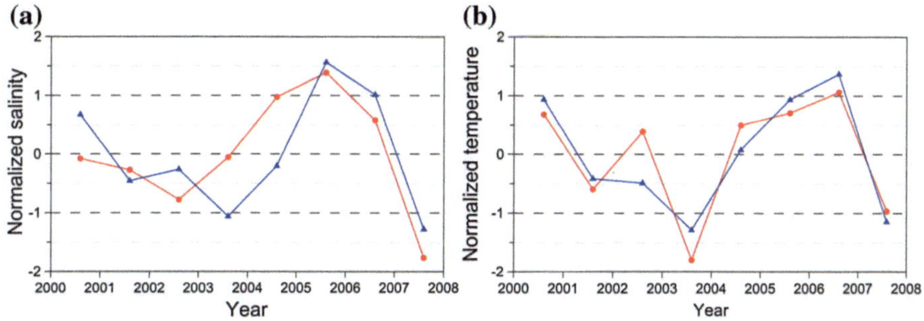

Fig. 4.15 Normalized values of salinity (**a**) and temperature (**b**) of the AW layer. The values from the southern study area are in *red*, from the northern one—in *blue*

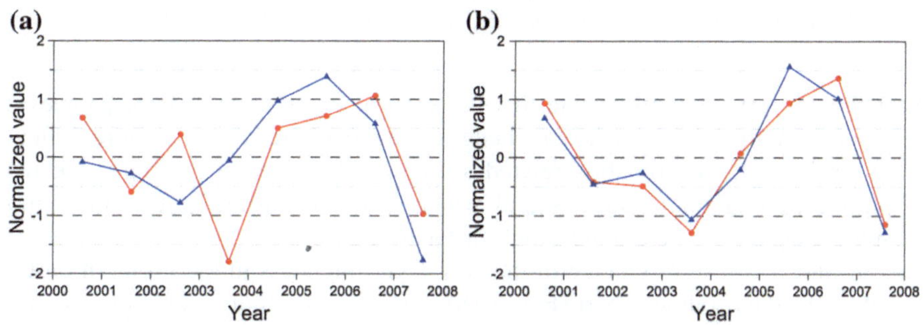

Fig. 4.16 Normalized values of salinity (*blue line*) and temperature (*red line*) of the AW layer in the southern (**a**) and northern (**b**) study area

was found for salinity, however, the correlation is lower. In the period 2000–2004 the salinity variations in the southern part appeared faster than in the northern one.

2. A lag in the temperature signal as compared to the salinity signal in the southern part was observed (higher correlation for 1-year time lag than for concurrent data, Fig. 4.16).
3. Temperature in the northern part correlates well with salinity in that part without any time lag.
4. There is a high correlation (r = **0.79**) between the AW layer salinity in the southern study area and the temperature in the northern study area with a time lag of one year (Fig. 4.17). It means that on the basis of the upstream measurements of the AW layer salinity in the southern study area, temperature change in the northern study area could be forecasted with high probability. If the relationship is confirmed in further studies, it will be possible to forecast a change in the temperature of the AW flowing into the Arctic Ocean through the Fram Strait with an advance of 1 year.

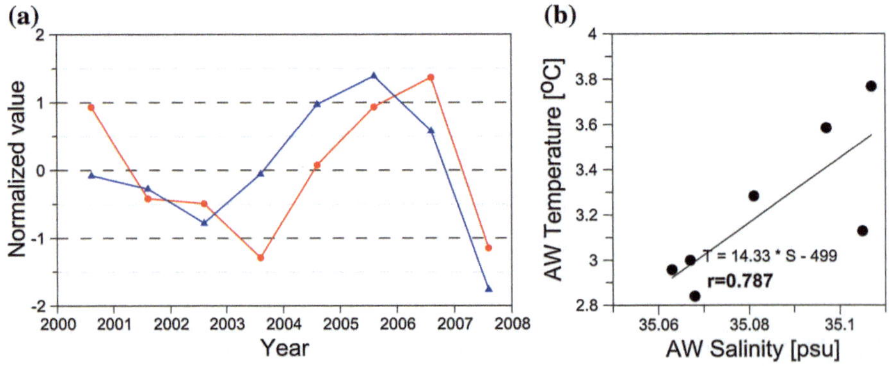

Fig. 4.17 Normalized, detrended values of salinity in the southern study area (*blue line*) and temperature in the northern study area (*red line*) of the AW layer (**a**), and regression salinity in the southern study area on temperature in the northern study area, shifted by 1 year (**b**)

5. Although it is comprehensible that the salinity signal appears in sequences in separate study areas, concurrent changes in temperature are much more difficult to explain. It is probably related to physical processes during advective propagation of temperature signal when, water, releasing heat to the atmosphere, changes temperature much faster than salinity.
6. The time advance of change in the AW layer thickness in the northern part in relation to the southern part (Fig. 4.18), opposite to salinity, is significant. The result seems coherent with previous considerations as regards the nature of the AW transport through the Fram Strait.

The change in the AW transport through the FS should firstly modify the AW layer thickness in the northern part and then in the southern one. However, the result

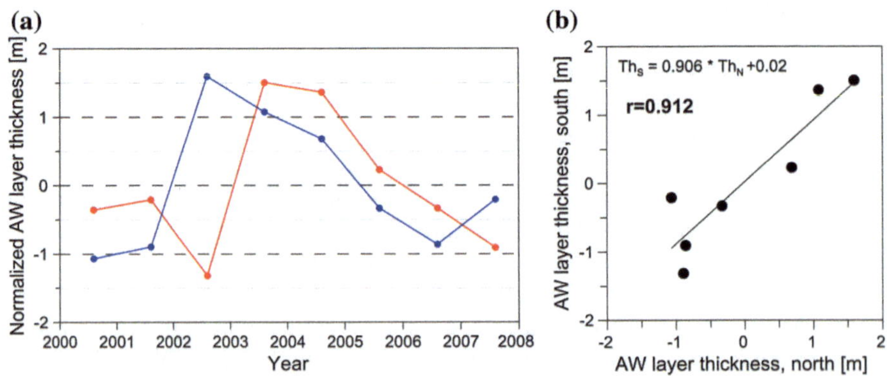

Fig. 4.18 Normalized detrended values of the AW thickness (**a**). The values from the southern study area are in *red*, from the northern one—in *blue*. Regression of standardized layer thickness in the northern and the southern study area, shifted in time by 1 year (**b**)

seems problematic—will detrending of the AW layer thickness data introduce a bias due to opposite trends in that case? The correlations are very low for raw data.

4.5 Correlation with the NAO index

While looking for the teleconnections, a decision was made to study the Atlantic Water properties with reference to large-scale atmospheric circulation. The winter NAO index was selected from among different indices.

Assuming that the AW in the Nordic Seas partially stores and transports the properties acquired in its formation region, the intensity of winds over the Atlantic Ocean may have impact not only upon the dynamics of the sea region but also—with a relevant time lag—upon the AW properties in the Norwegian or the Greenland Sea. The NAO index values have been downloaded from http://www.cgd.ucar.edu/cas/jhurrell/indices.html (Hurrell et al. 2003).

The potential of combining one index with variety of dynamical processes is very tempting for researchers dealing with various processes. Numerous papers describe correlations between the NAO and temperature and salinity in the North Atlantic and the Nordic Seas (Blindheim et al. 2000; Polyakov et al. 2003; Bersch et al. 2007), the intensity of deep convection in the Labrador Sea and the Greenland Sea (Lilly et al. 1999), the area and concentration of ice transported through the Fram Strait (Divine and Dick 2006; Koenigk et al. 2006; Kinnard et al. 2008), the intensity of the AW inflow into the Arctic Ocean and its temperature (Dickson et al. 2000). Tables 4.6, 4.7 and 4.8 show the correlation coefficients between mean AW properties and the NAO index for the entire study area, the southern part and the northern part.

The correlation coefficients *r* were calculated first for the 2000–2006 time series, and then (after the 2007 data were obtained) for the 2000–2007 series. No correlations between mean AW properties and the winter NAO index in the given year or in the previous year were found. There were higher correlations between the AW properties and the NAO index with a backward time lag of 2–4 years. In most cases the correlation is negative. Correlations for the period 2000–2006 are much higher, while a sudden decrease of the AW temperature and salinity in 2007 resulted in lower correlations for the longer time series, however, the conclusions

Table 4.6 Coefficients of correlation *r* between the winter NAO index and selected properties of the AW from the entire AREX study area

2000–2006 parameter	Time lag (years)	*r*	2000–2007 parameter	Time lag (years)	*r*
S	4	**−0.78**	S	4	−0.31
T	4	−0.66	T	4	−0.42
T	3	−0.63	T	3	−0.47
Q	3	**−0.79**	Q	3	−0.21
Th	4	0.62	Th	4	0.66

Table 4.7 Coefficients of correlation *r* between the winter NAO index and selected properties of the AW in the southern part of the study area

2000–2006 parameter	Time lag (years)	*r*	2000–2007 parameter	Time lag (years)	*r*
S	3	−0.61	S	0	−0.49
T	3	−0.61	T	3	−0.52
Q	3	**−0.89**	Q	3	−0.55
Th	2	−0.66	Th	2	−0.63

Table 4.8 Coefficients of correlation *r* between the winter NAO index and selected properties of the AW in the northern part of the study area

2000–2006 parameter	Time lag (years)	*r*	2000–2007 parameter	Time lag (years)	*r*
S	4	−0.84	S	4	−0.52
T	4	−0.70	T	4	−0.47
T	3	−0.54	T	3	−0.41
Q	4	−0.71	Q	4	−0.38
Th	4	0.61	Th	4	0.62

as regards the time lag remained the same. Mean properties of the AW respond to the NAO change with a lag of 2–4 years. The correlation between the NAO and the AW layer thickness is the most robust. In the southern part it is negative, with a time lag of 2 years, and in the northern part the time lag rises to 4 years, however, the correlation coefficient becomes positive.

The lack of correlation without a time lag is interesting and quite unexpected. There are some papers which relate higher temperature of the upper AW layer flowing through the eastern part of the Nordic Seas (Furevik 2000) or the intense inflow of the warmer AW into the Nordic Seas and the Arctic Ocean (Dickson et al. 2000; Bersh 2004; Furevik and Nilsen 2005) to the positive NAO index. Also on the basis of the IOPAS historical data, Schlichtholz and Goszczko (2006) found some significant correlations in the period 1991–2003 between the winter NAO index and mean AW properties at the latitude of 76°30′ (section "N") without a time lag.

The reason for the differences may be the fact that the influence of the NAO index on the SLP is not fixed in time; correlations of the NAO index with a parameter may break in the event that the NAO's spatial structure changes (Furevik and Nilsen 2005). Another, less probable cause is applying different definitions of the AW.

It is meaningful that in the event of correlations with a time lag, one can see a logical sequence of ocean climate changes occurring in space, caused by changes in the NAO index. In the southern part the highest correlations occur with a time lag of 3 years, and in the northern one with a lag of 4 years. It suggests that the correlations are not random, the change in the AW properties was not local and the signal advection may be observed. A positive NAO phase causes temperature and salinity to decrease in the study area after 3–4 years, whereas a negative one causes an increase. Significant time lags between the NAO index and the dynamical processes in the Nordic Seas are also known from various reference books. Orvik and Skagseth (2003) described a time lag of 15 months between the change in wind vorticity above the North Atlantic along the 55°N parallel and the change in the flow at the Svinoy Section. Blindheim et al. (2000) negatively correlated the spatial extension of the Atlantic waters with the NAO index shifted by more than 3 years backwards, explaining the same by a change in the structure of the AW inflow into the Nordic Seas. Nilsen et al. (2003) found a negative correlation between two AW branches flowing into the Nordic Seas—the Norwegian-Atlantic Current (western branch) and the Norwegian-Atlantic Slope Current (eastern branch), and they related this feature to changes in the NAO index.

4.6 Correlation with Temperature in Hornsund

In order to study the relationship between the AW temperature and mean atmospheric conditions in the region locally influenced by the warm AW inflow, a series of air temperature data gathered by the Polish Polar Station in Hornsund was used. Only the monthly means (Fig. 4.19) and annual means were employed for the purpose of the present work. Although the Hornsund Fjord located on the south-eastern tip of Svalbard is additionally influenced by the cold Sørkapp Current (Marsz and Styszyńska 2007), the West Spitsbergen Current is the most important current in the region.

Similarly, the atmospheric circulation in that area is shaped mostly by the processes occurring in the Atlantic Arctic (Niedźwiedź 2007). The oceanic climate index is stronger than for other weather stations on Svalbard and oceanic air masses occur in Horsund for 85 % of days per year (Marsz 2007).

Mean monthly temperature in Hornsund indicates an annual cycle of ca. 18 °C amplitude and a positive trend. Mean temperature in July and August is above 4 °C, characterized by low variability and shows no distinct trend, however, in the coldest winter months (January, February, March), mean temperature falls below

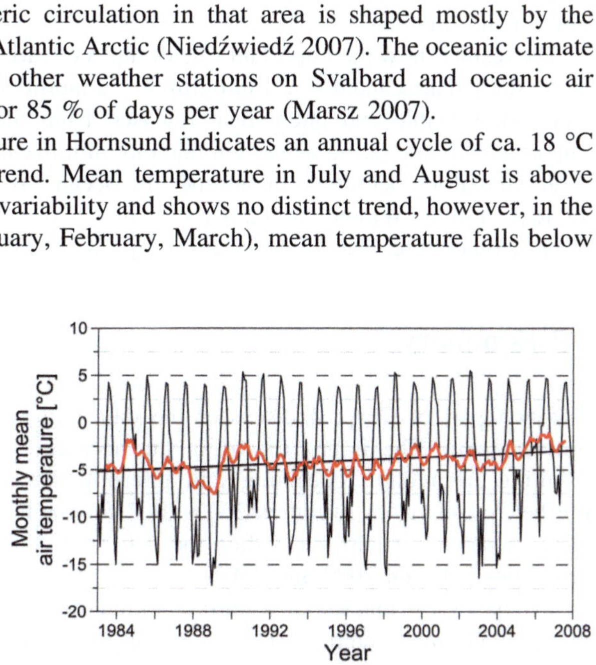

Fig. 4.19 A series of mean monthly air temperatures in the Polish Polar Station in Hornsund. The linear trend is marked by the *black line* and the running mean of 11 months' period is marked by the *red line*

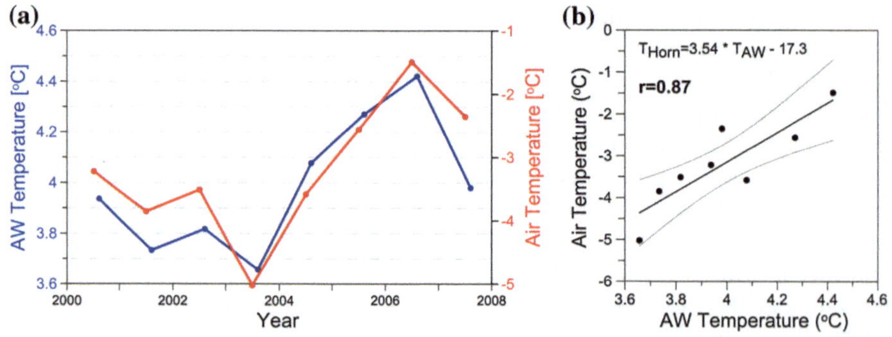

Fig. 4.20 A time series of mean annual air temperature in the Polish Polar Station in Hornsund (*red line*) and mean Atlantic Water temperature in the AREX study area in July (*blue line*) (**a**); regression of the both values (**b**). The 95 % confidence level is marked

−10 °C, indicating high interannual variability and a significant positive trend. Very low temperature in winter 2003 and an increase of winter temperature since 2004 is characteristic.

The similarity between the values of mean AW temperature in the entire AREX study area in July and the mean annual air temperature in Hornsund is high (Fig. 4.20). The minimum in 2003 and maximum in 2006 are clearly indicated.

The coefficient of correlation between the values of mean annual air temperature in Hornsund and mean Atlantic Water temperature in the AREX study area is **0.86**. For detrended data the correlation coefficient slightly decreases to **0.78**. There is no time lag.

The results show the important role of the AW variability in shaping the climate in the studied area and how meaningful the variability of the AW properties is. The fluxes of sensible and latent heat released from the ocean are of the highest importance in mitigating winter temperature minima and that is how they influence mean annual temperatures. In summer the ocean has a negligible effect upon the air temperature. Winter air temperature has also a substantial impact upon the ocean temperature. The above relationships will be discussed again in Chap. 5.

4.7 Summary

The first part of Chap. 4 discusses the data collection and methods applied to process them. The issues regarding the accuracy of measurements and errors resulting from data processing and interpolation were also briefly discussed. Further, means from summers 2000–2007 for the temperature and salinity fields were presented. Then, the author focused on time variability of the AW layer properties between 2000 and 2007.

The data show that initially, until 2003, there was a period of moderate decrease of the AW layer temperature and salinity, and after that a period of sudden warming and increase of salinity occurred. The increase of temperature also caused the increase of heat content in the AW layer. In 2007 there was a sudden decrease of temperature and salinity. HIgh correlations were found between mean temperature and salinity of the AW in the entire study area and temperature, salinity and heat content in the AW layer. No correlations were found between the basic AW properties and the layer thickness in a given year. Another interesting result is a negative correlation between temperature (and salinity) of the AW layer and its thickness in the following year.

The analysis of the AW properties with use of the Empirical Orthogonal Functions showed that the first EOF mode described more than 60 % of the fields' variance.

Then, due to observed differences in the spatial structure of hydrographic fields, a suggestion was made to divide the study area into two parts—the northern one and the southern one, with a boundary along the 74°N parallel. The analysis of data from each part separately proved that the salinity signal propagates in a sequence, a change occurs first in the southern part and then in the north. Mean salinity in the southern part in the period 2000–2006 is highly correlated with mean temperature in the northern part in 2001–2007 period.

The potential link between variability of AW properties and external atmospheric forcing was also studied, which was done by analyzing correlations between the AW properties and the winter NAO index. No significant correlation between the NAO and temperature or salinity in the same year was found. Higher correlations (mostly negative) were obtained with a time lag of 3-4 years, however, the majority of them were below the 95 % level of confidence. One can very generally state that negative correlations were found (the lower the NAO index—the higher the salinity or temperature) with ca. 2–3 years' time lag in the southern study area, and ca. 3–4 years' lag in the northern part. It also proves that the properties of the AW in the Nordic Seas are influenced by conditions prevailing in the region where the AW is formed, however, it is strongly modified during its northward advection.

The end of the chapter describes the impact of the AW properties variability upon the climate in the selected region of the Atlantic Arctic. A very strong correlation was found between the AW mean temperature and mean annual temperature measured at the Polish Polar Station in Hornsund.

References

Bersch M (2004) North Atlantic Oscillation–induced changes of the upper layer circulation in the northern North Atlantic Ocean. J Geophys Res 107(C10):3156, doi:10.1029/2001JC000901

Bersch M, Yashayaev I, Koltermann KP (2007) Recent changes of the thermohaline circulation in the subpolar North Atlantic. Ocean Dyn 57:223–235. doi:10.1007/s10236-007-0104-7

Blindheim J, Borovkov V, Hansen B, Malmberg S-Aa, Turrell WR, Østerhus S (2000) Upper layer cooling and freshening in the Norwegian Sea in relation to atmospheric forcing. Deep-Sea Res I 47:655–680

Dickson RR, Osborn TJ, Hurrel JW, Meincke J, Blindheim J, Adlandsvik B, Vinje T, Alekseev G, Maslowski W (2000) The arctic ocean response to the North Atlantic Oscillation. J Clim 13:2671–2696

Divine D, Dick C (2006) Historical variability of sea ice edge position in the Nordic Seas. J Geophys Res 111:C01001. doi:10.1029/2004JC002851

Dukhovskoy DS, Johnson MA, Proshutinsky A (2004) Arctic decadal variability: an auto-oscillatory system of heat and freshwater exchange. Geophys Res Lett 31:L03302. doi: 10.1029/2003GL019023

Dukhovskoy D, Johnson M, Proshutinsky A (2006a) Arctic decadal variability from an idealized atmosphere-ice-ocean model: 1. Model description, calibration, and validation. J Geophys Res 111:C06028. doi:10.1029/2004JC002821

Dukhovskoy D, Johnson M, Proshutinsky A (2006b) Arctic decadal variability from an idealized atmosphere-ice-ocean model: 2. Simulation of decadal oscillations. J Geophys Res 111: C06029. doi:10.1029/2004JC002820

Furevik T (2000) On anomalous sea surface temperatures in the Nordic Seas. J Clim 135: 1044–1053

Furevik T, Nilsen J (2005) Large-scale atmospheric circulation variability and its impacts on the nordic seas ocean climate—a Review. In Drange et al (eds) The Nordic Seas: an Integrated Perspective, Geophysical Monograph Series, 158, AGU

Hopkins TS (1991) The GIN Sea—A synthesis of its physical oceanography and literature review 1972–1985. Earth Sci Rev 30:175–318

Hurrell JW, Kushnir Y, Ottersen G, Visbeck M (2003) An overview of the north atlantic oscillation. The north atlantic oscillation: climatic significance and environmental impact. Geophysical Monograph 134, American Geophysical Union, doi:10.1029/134GM01

Kinnard C, Zdanowicz CM, Koerner RM, Fisher DA (2008) A changing Arctic seasonal ice zone: observations from 1870–2003 and possible oceanographic consequences. Geophys Res Lett 35:L02507. doi:10.1029/2007GL032507

Koenigk T, Mikolajewicz U, Haak H, Jungclaus J (2006) Variability of Fram Strait sea ice export: causes, impacts and feedbacks in a coupled climate model. Clim Dyn 26:17–34. doi:10.1007/s00382-005-0060-1

Lilly J, Rhines P, Visbeck M, Davis R, Lazier J, Schott F, Farmer D (1999) Observing Deep Convection in the Labrador Sea during Winter 1994/95. J Phys Oceanogr 29:2065–2098

Manley TO (1995) Branching of Atlantic Water within the Greenland-Spitsbergen Passage: an estimate of recirculation. J Geophys Res 100(C10):20627–20634

Marsz A (2007) Klimat stacji w świetle niektórych wskaźników klimatycznych. Klimat Polskiej stacji polarnej w Hornsundzie, Akademia Morska w Gdyni, p 376

Marsz A, Styszyńska A (2007) Główne czynniki klimatyczne. Klimat Polskiej stacji polarnej w Hornsundzie, Akademia Morska w Gdyni, p 376

Niedźwiedź T (2007) Cyrkulacja atmosferyczna. Klimat Polskiej stacji polarnej w Hornsundzie, Akademia Morska w Gdyni, p 376

Nilsen JEØ, Gao Y, Drange H, Furevik T, Bentsen M (2003) Simulated North Atlantic–Nordic Seas water mass exchanges in an isopycnic coordinate OGCM. Geophys Res Lett 30(10): 1536. doi:10.1029/2002GL016597

Orvik KA, Skagseth Ø (2003) The impact of the wind stress curl in the North Atlantic on the Atlantic inflow to the Norwegian Sea toward the Arctic. Geophys Res Lett 30(17):1884. doi:10.1029/2003GL017932

Polyakov IV, Bekryaev RV, Bhatt US, Colony RL, Maskshtas AP, Walsh D, Bekryaev RV, Alekseev GV, Johnson MA (2003) Variability and trends of air temperature and pressure in the maritime Arctic, 1875–2000. J Clim 16:2067–2077

Schlichtholz P, Goszczko I (2006) Interannual variability of the Atlantic water layer in the West Spitsbergen Current at 76,5°N in summer 1991–2003. Deep-Sea Res I, 53, doi:10.1016/j

Walczowski W (1997) Transfrontalna wymiana masy i ciepła w rejonie Frontu Arktycznego, PhD Thesis, IOPAN, p 123

Chapter 5
Structure and Variability of the Atlantic Water on the Transects

The Atlantic Water can be easily distinguished from other water masses occurring in the investigated region. It is characterized mostly by much higher temperature and salinity. The higher temperature is a factor which reduces the water density; the higher salinity increases the density. For the AW, temperature is the prevailing factor—it makes the AW less dense than the surrounding waters. It results from high ratio between the thermal expansion coefficient α:

$$\alpha = \frac{1}{\rho}\left(\frac{\partial \rho}{\partial T}\right)_{p,S} \tag{5.1}$$

and the salinity contraction coefficient β (modification of volume due to change in salinity):

$$\beta = \frac{1}{\rho}\left(\frac{\partial \rho}{\partial S}\right)_{p,T} \tag{5.2}$$

where ρ is water density. S and T are salinity and temperature respectively. The result of the above is the salinity inversion—the more saline AW lays above less saline and colder intermediate and deep waters (Fig. 5.1). Such stratification makes the column more susceptible to convection (Druet 1994).

Figure 5.1 shows water column structure at several stations in the centre of the Atlantic Domain where the inflow of the Arctic and Polar Water is negligible, and closer to the Arctic Front where the AW is colder and less saline. The AW layer thickness varies from 200 to 700 m. The warmest AW occupies the surface layer and it is bounded by the summer thermocline at 60 m. The proper thermocline and halocline which separates the AW layer from the intermediate water lies at the depth of ca. 600 m. Below that depth there are stratified intermediate waters and homogenous deep waters. At the profiles which are situated closer to the domain boundary, the halocline and thermocline are less distinct and at smaller depths.

Collecting the data along hydrographic transects (sections) is a standard method of oceanographic survey. Transects are most often located across straits, perpendicularly to the direction of currents and isobaths. On the basis of previous studies

W. Walczowski, *Atlantic Water in the Nordic Seas*,
GeoPlanet: Earth and Planetary Sciences, DOI: 10.1007/978-3-319-01279-7_5,

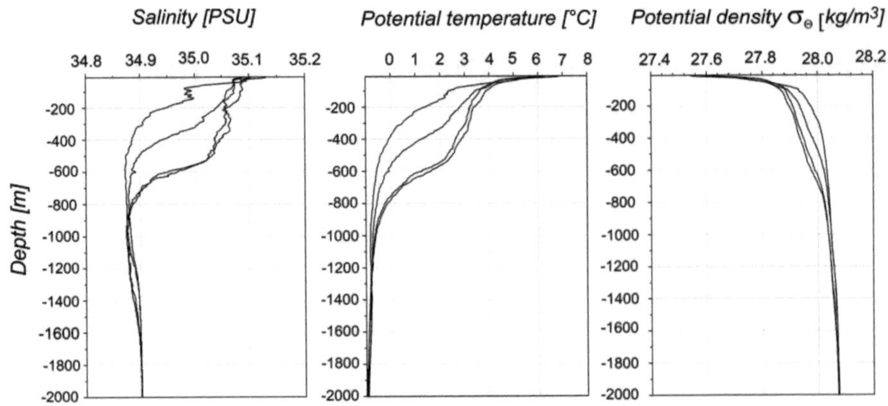

Fig. 5.1 Salinity (PSU), temperature (°C) and potential density (kg/m^3) profiles taken on transect 'K' in 2001

of the WSC, the majority of transects was performed perpendicularly to the presumed direction of current. A series of stations repeated year by year made it possible to measure horizontal fields of temperature and salinity which picture the spatial distribution and properties of water masses well. The stations are arranged along transects, which provides a better representation of vertical distribution of water masses, their structure and mutual relations. Furthermore, time series of the AW properties at individual transects allow for some (limited) tracing of signal.

5.1 Transect '*H*'

Transect H runs along the parallel 73°30′N. According to the previously introduced division, it covers the northern section of the southern part of the study area. The transect is situated in the northern part of the main area where the AW flows into the Barents Sea through the BSO. In the western part of the transect, meridional pattern of all the isolines prevails while fluxes into the Barents Sea dominate in the eastern part (see horizontal distributions). An exemplary section from 2004 shows the water mass structure at the transect 'H' (Fig. 5.2). As in all other zonal transects, west is to the left. The transect covers the Atlantic Domain from the Arctic Front to the Barents Sea shelf. The lower AW boundary is marked with bold isolines: 0 °C isotherm and 34.92 isohaline. The AW vertical extent is most frequently limited by the 34.92 isohaline, since the 0 °C isotherm is usually situated below that line. The AW extends over a width of more than 450 km, covering 202 km^2 of the transect 'H'. The AW mean temperature is 3.96 °C and its mean salinity is 35.07.

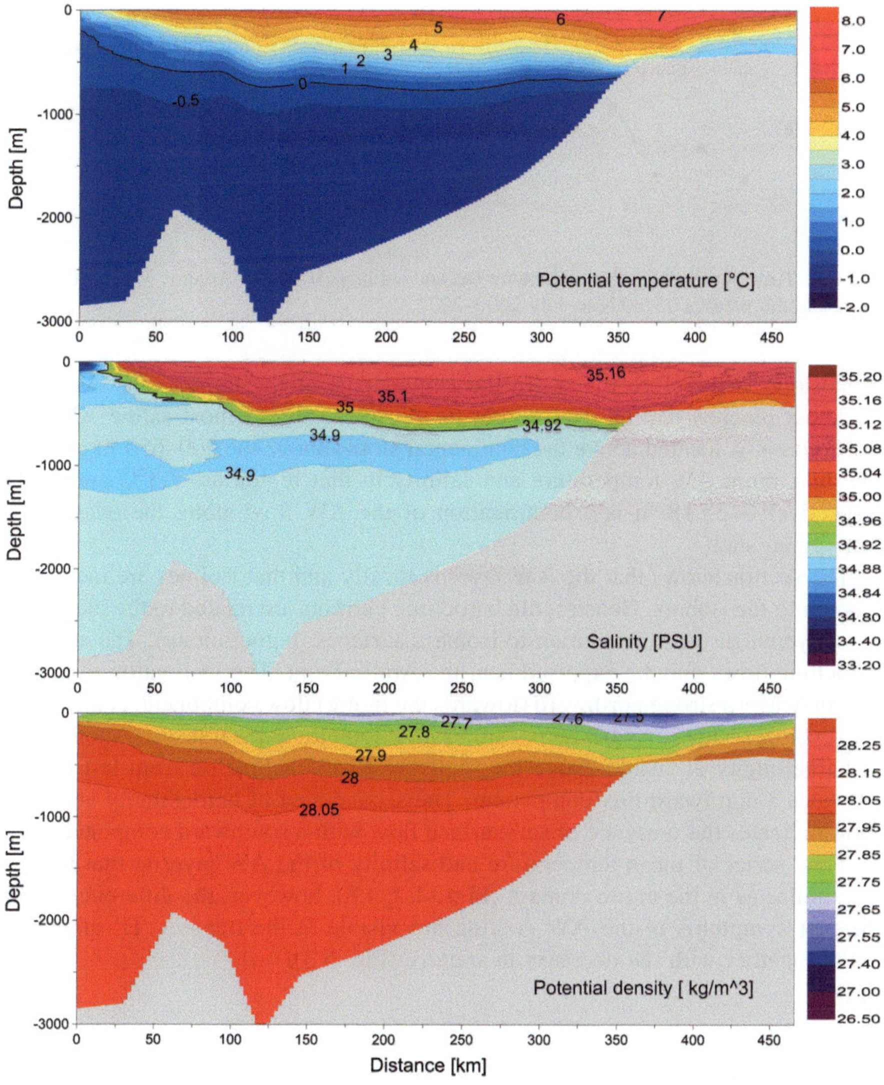

Fig. 5.2 Temperature, salinity and potential density on transect 'H' along the parallel 73°30′N. July 2004, r/v Oceania. The 0 °C isotherm and 34.92 isohaline are bold

The impact of the bottom topography upon the distributions of hydrographic properties is clearly distinguished at both ends of the transect. In the western part the location of the Arctic Front, dividing the Atlantic Water and the Arctic Water coincides with a line of underwater ridges. The front is recognized by the inclination of isotherms, isohalines and isopycnals in relation to isobaric surfaces. Due to differences in horizontal and vertical scale, the front slope on the figure is excessive. Along a distance of 100 km the isolines sink by less than 500 m; the

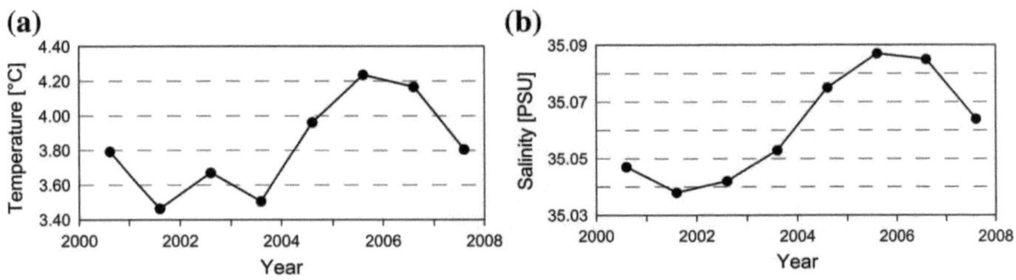

Fig. 5.3 Time series of mean temperature (**a**) and salinity (**b**) of the Atlantic Water at transect 'H' along the parallel 73°30′N in July 2000–2007

front slope is therefore 0.5 %. In the central part of the transect the AW layer thickness reaches 650 m. The region of the warmest and most saline AW—the WSC core—is located above the continental slope, above the 500–650 m isobaths. The maximum AW temperature and salinity in that region are 7.5°C and 35.18, respectively—35.18. It is a continuation of the AW flow along the edge of the Norwegian shelf.

The section shows that the AW layer is locally and that isolines are inclined in relation to the isobars. Geostrophic baroclinic currents are related to the inclination of isopycnal surfaces in relation to isobaric surfaces (baroclinicity). The strongest baroclinic flows may be expected near the Arctic Front. The inclination of isopycnals indicates a strong northward (towards the figure) flow component. A mesoscale anti-cyclonic eddy is visible above the 125th kilometre of the transect. Another area of baroclinicity is visible above the shelf break. Its inclination from left to right indicates a northward flux component. The second area of baroclinicity above the shelf indicates the presence of sub-surface flow with a southward component.

Time series of mean temperature and salinity of the AW layer at that transect are similar as in the entire domain (Figs. 4.7, 4.8), however, the difference is that the first symptoms of the AW cooling are visible at the transect 'H' already in 2006, together with the decrease in salinity (Fig. 5.3).

5.2 Transect '*K*'

The exemplary transect along the parallel 75°N, from the longitude 05 to 17°E, is shown in Fig. 5.4. The reason why the limit values of T > 0 °C and S > 34.92 were chosen for definition of the AW layer is clearly visible. These are the ranges of the strongest vertical gradients of those values, separating intermediate water from the AW layer. At that transect the AW covers the width of approximately 340 km and the area of 145 km^2. The AW mean temperature is 3.08 °C and mean salinity is 35.02. In the centre of the domain the AW layer thickness reaches values between 500 and 700 m. The AW layer is thicker on the western side, near the Arctic Front, and in the east, above the shelf slope of the Barents Sea. These

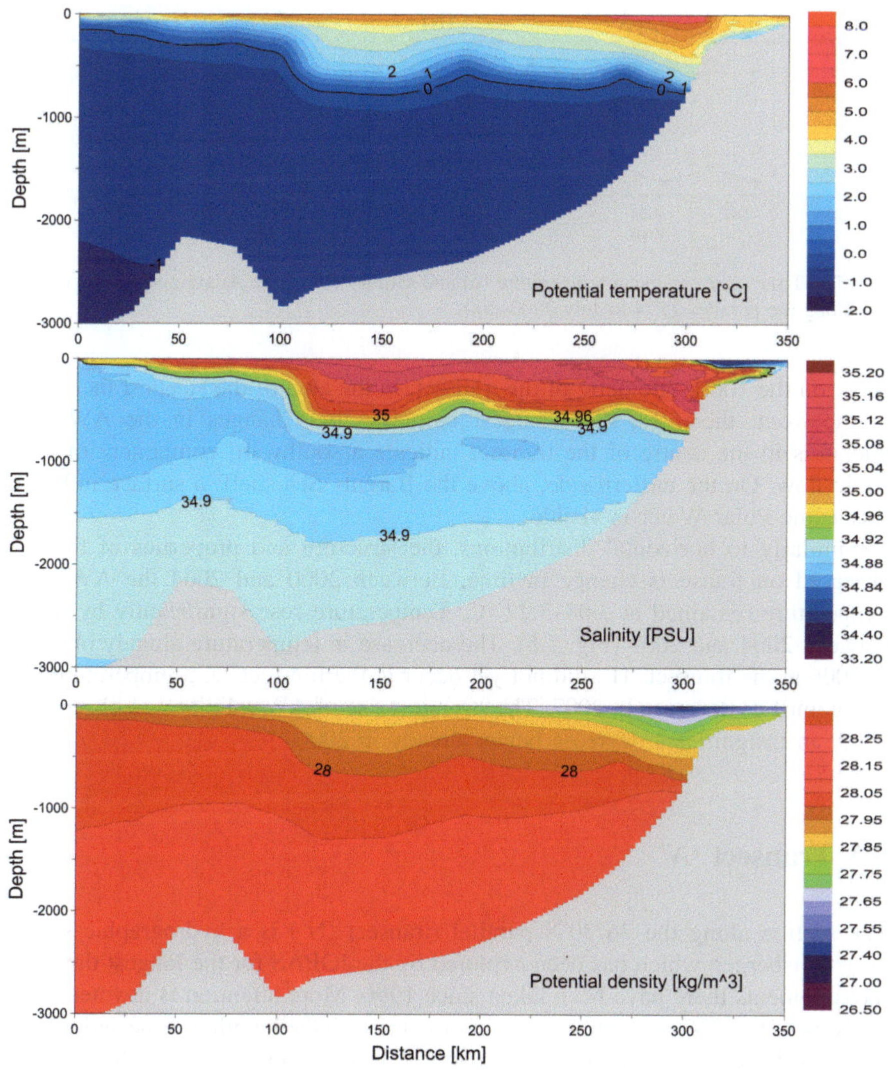

Fig. 5.4 Temperature, salinity and potential density on transect 'K' along the parallel 75°N. July 2001, r/v Oceania. The 0 °C isotherm and 34.92 isohaline are bold

maxima represent two branches of the West Spitsbergen Current. The warmest and most saline water flows above the slope, along the 600–800 m isobaths. This transects also features a bottom topography—including in particular the underwater ridge situated on the 50th kilometre—which limits the Atlantic Domain and reflects the location of the Arctic Front. The jet streams, generated by the difference in the water density on both sides of the front and strong baroclinicity, are connected with the front. In the discussed case the most intense currents should

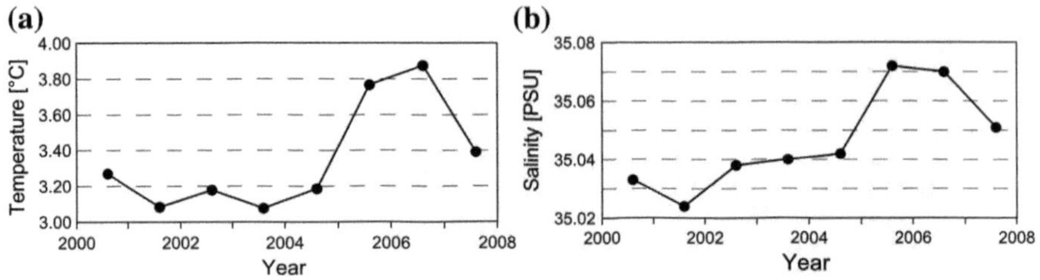

Fig. 5.5 Time series of mean temperature (**a**) and salinity (**b**) of the Atlantic Water on transect 'K' along the parallel 75°N in July 2000–2007

occur on the 100th kilometre of the transect, in the area of the steepest inclination of the front, therefore, the strongest baroclinicity. Changes in the AW layer thickness in the centre of the transect indicate a southward component of baroclinic flow. On the eastern side, above the Barents Sea shelf, a surface inflow of less saline Polar Water is visible.

Similarly to horizontal distributions, the structure and properties of the AW observed on transects change in time. Between 2000 and 2004 the AW mean temperature remained at 3.08–3.27 °C. Temperature rose significantly by 0.6 °C between 2004 and 2005 (Fig. 5.5). The decrease in temperature already observed in 2006 at the transect 'H', did not yet occur at the transect 'K'; temperature kept rising until its decrease in 2007. The strongest rise of AW salinity was observed in 2005, an insignificant decrease began already in 2006.

5.3 Transect '*N*'

The section along the 76°30′N parallel (transect 'N') is a hydrographic section near Spitsbergen which has been explored by the IOPAS for the longest time. The measurements there have been taken since 1996. More attention is devoted to the analysis of the AW properties at that transect due to the length of time series and a special role of that sea region. It is a sort of gateway into the Fram Strait where different WSC branches converge. On transects further up north, the currents may not be so easy to distinguish.

The coverage of the western part with measurements varied in different years, mostly due to variable ice conditions. Since 2000, the transect has been measured between 004°E and the Spitsbergen shelf, and it covered the entire Atlantic Domain.

Similarly as in the case of the transect 'K', the AW spreads at a width of approximately 340 km (Fig. 5.6). However, the proper width of the Atlantic Domain is smaller, approximately. 280 km, and its average area is 143 km^2. The transition zone between the AD and the Arctic water is the Arctic Front which begins at the 60th kilometre of the transect; the Atlantic Water west of the AF recirculates into the Greenland Sea. The AW layer thickness reaches 670 m.

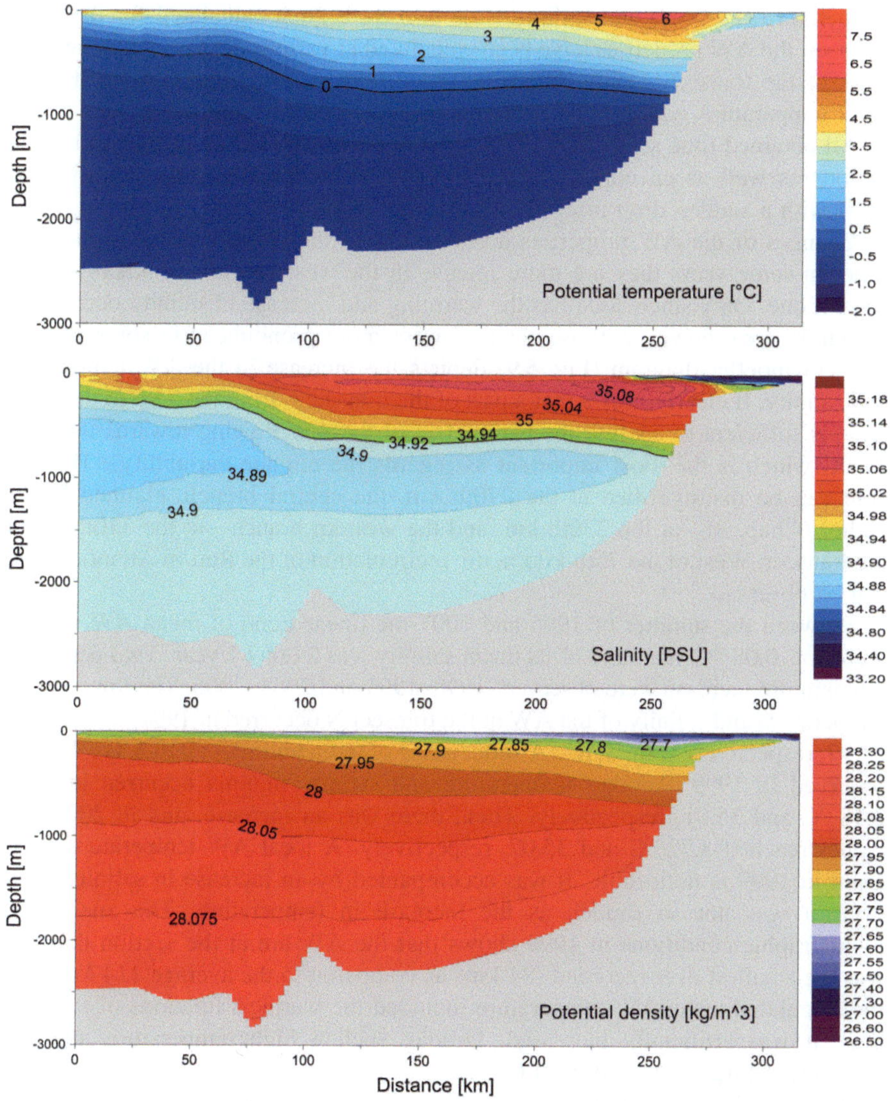

Fig. 5.6 Potential temperature, salinity and potential density on transect 'N' along the parallel 76°30′N. 1996–2007 summer mean, data gathered aboard r/v Oceania. The 0 °C isotherm and 34.92 isohaline are bold

Similarly to the previous transects, two zones of baroclinicity and resulting baroclinic currents are clearly distinct: above the Knipovich Ridge (the western WSC branch) and above the Spitsbergen slope (the WSC core).

Temporal variability of the AW properties along that transect has been analysed and presented in different ways. Mean values (Fig. 5.6) as well as anomalies of temperature (Fig. 5.7) and salinity (Fig. 5.8) have been calculated for the entire transect; the AW properties have been calculated at various depths and for selected parts of the transect. Finally, the time-space (Hovmoeller) diagrams of the AW layer temperature, salinity and heat content have been presented.

All obtained time series of derived parameters clearly show warmer and colder periods as well as an unprecedented increase in temperature and salinity since 2004 with a sudden drop in 2007.

Changes of the AW properties do not occur at the entire transect at the same time. In some years they are more intense in the western part, in others—in the eastern one. Only since 2004 has the warming and increase in salinity occurred in the entire area, however, those changes were more pronounced in some regions. The Hovmoeller diagram (Fig. 5.9) depicts the increase in the AW temperature and salinity. It clearly shows that a part of the transect—from the 75th to the 275th km—is sufficient to analyse the parameters of the AW flowing towards the Fram Strait (which is the most important as regards the climate variability). The AW core can be distinguished at the 270th km, the central branch, claimed by the author (Chap. 8)—at the 230th km, and the western branch—at the 110th km of the transect. West of the 75th kilometre, recirculation of the Return Atlantic Water can be observed.

Between the summer of 1996 and 2007 the linear trend of mean AW temperature was 0.08 °C/year, and of its mean salinity was 0.0067 1/year. Two periods of temperature increase were observed: 1998–1999 and 2004–2006. The lowest mean temperature and salinity of the AW at the transect N occurred in 1997: 2.07 °C and 34.97, respectively (Fig. 5.10). A sudden rise of temperature—up to 3.41 °C—was observed in 1999. Another temperature and salinity minima occurred in 2003: 2.48 °C and 35.01, respectively. Then, there was an increase, and in 2006 both values reached 3.72 °C and 35.07, respectively. A local AW temperature maximum in 1999 is noticeable. It was accompanied by an increase in salinity, however, it was not so sudden as the increase in temperature. The analysis of hydrographic conditions in 1999 shows that the AW are at the section that year was the smallest ever recorded (84 km^2 as compared to the average 114 km^2) and the calculated mean AW temperature included the warmest fractions of the flow, thus, it was artificially increased. Despite such a high temperature, the AW transported little heat in 1999 (Fig. 5.14).

Transect 'N' can be regarded as representative for the entire investigated area. In the period between 2000 and 2007 the correlation between the AW temperature at the transect and the mean temperature of the AW layer in the entire AREX study area equals **0.96**, and for salinity the correlation coefficient equals **0.83**.

Similarly as for the entire AW volume in the AREX study area, the AW temperature at the transect 'N' was related to air temperatures at the Polish Polar Station in Hornsund. The similarity between the time series of AW temperature and annual mean of air temperature in Hornsund is very high (Fig. 5.11).

The correlation coefficient between both values equals **0.92** and is even higher than for the entire AW mass. The result is even more valuable considering the fact that it has been obtained from a longer time series. For detrended data $r = \mathbf{0.86}$.

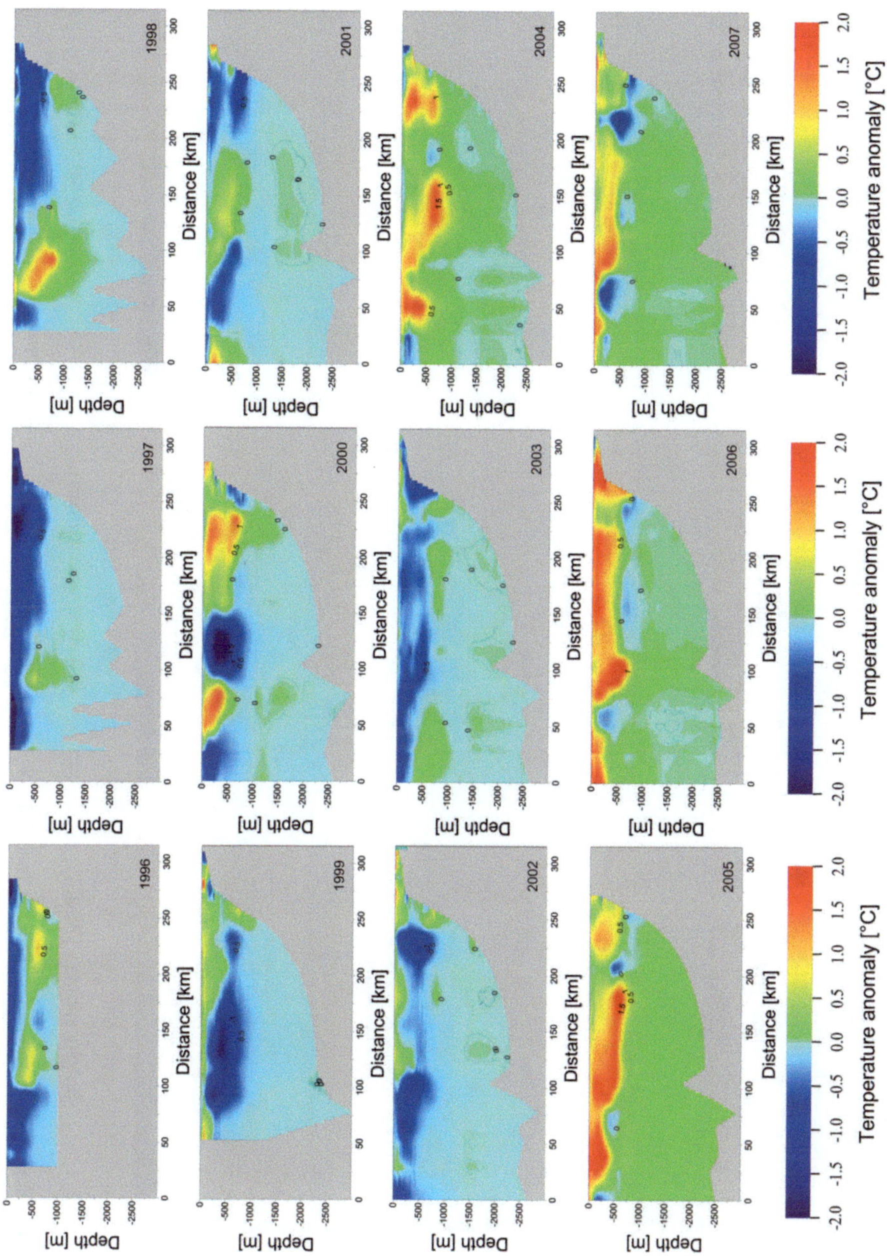

Fig. 5.7 Temperature anomalies (°C) in June–July on transect 'N', calculated in relation to the 1996–2007 mean

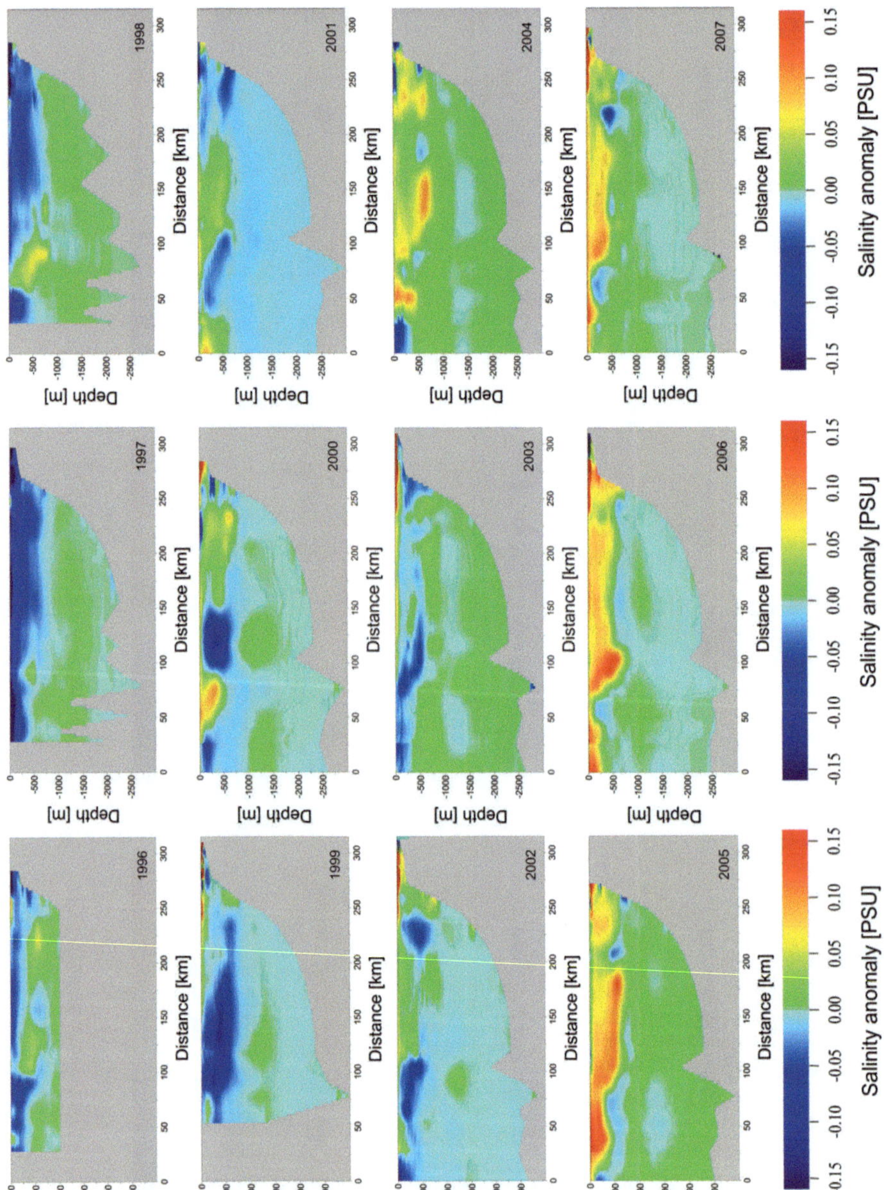

Fig. 5.8 Salinity anomalies (PSU) in June–July on transect ‘N’, calculated in relation to the 1996–2007 mean

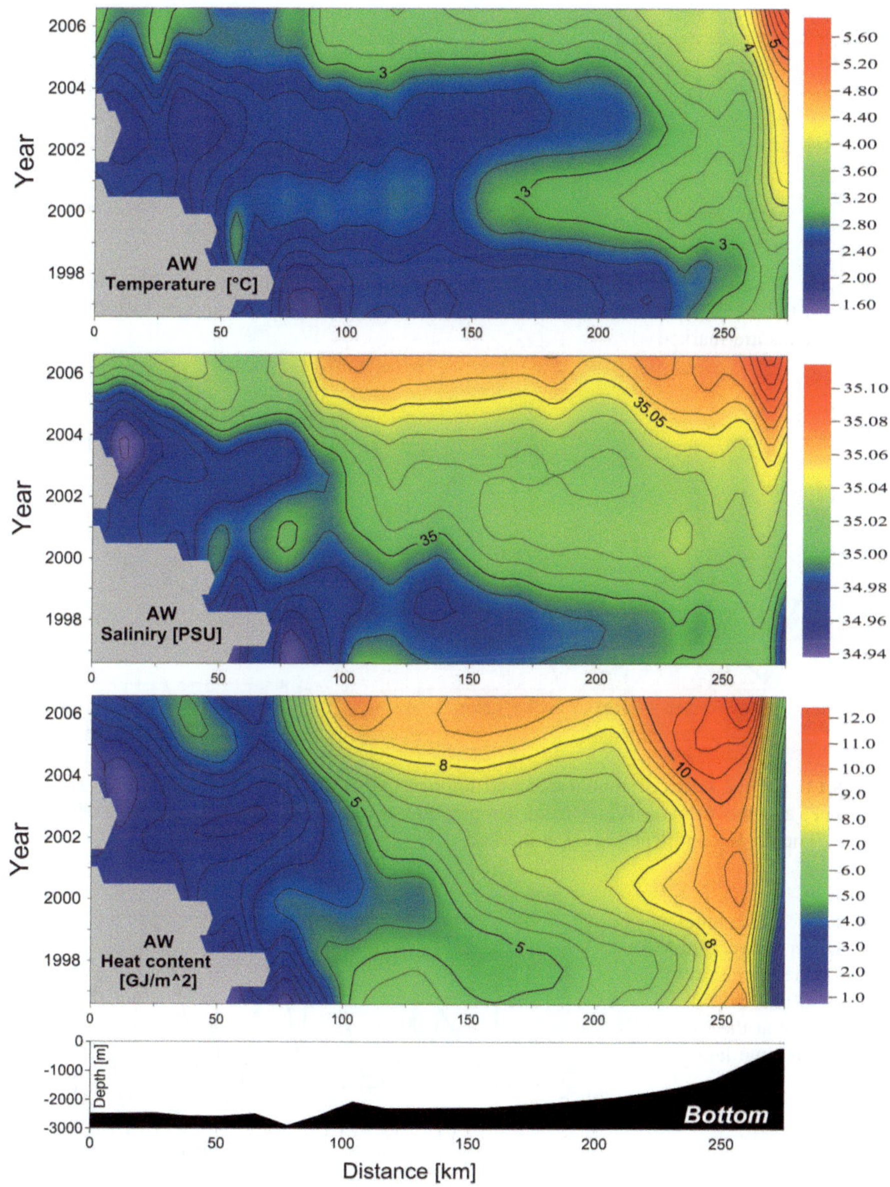

Fig. 5.9 Hovmoeller diagrams of temperature (°C), salinity (PSU) and heat content (GJ/m^2) of the AW layer on transect 'N' from the period 1996–2007. The *bottom* profile is shown. *Grey areas* mean lack of data or non-occurrence of the AW (in 1997)

Annual means of air temperature were calculated from monthly means, from January to December, the AW temperature was measured in July of the given year, in the middle of that period. Therefore, a question arises whether the AW

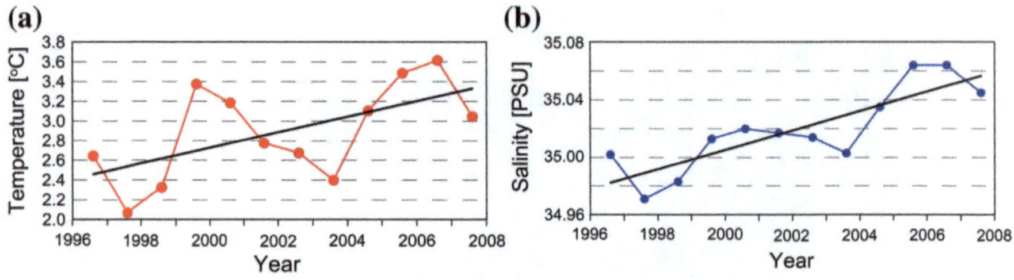

Fig. 5.10 Time series of mean temperature (**a**) and salinity (**b**) of the Atlantic Water on transect 'N' along the parallel 76°30′N, between the 75th and the 275th of the transect, July 1996–2007. Linear trends are marked

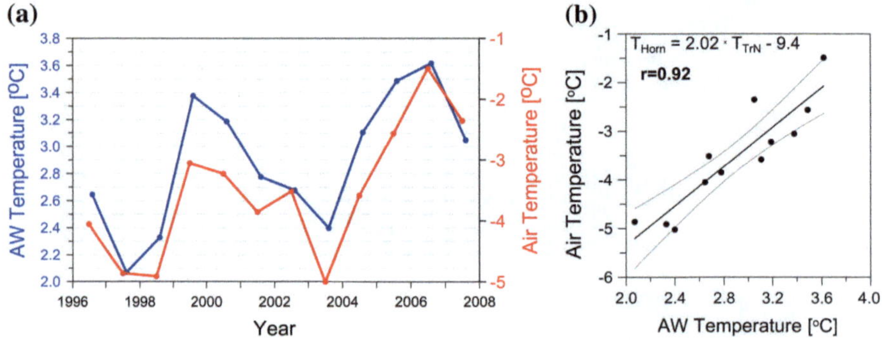

Fig. 5.11 **a** Time series of annual mean air temperature at the Polish Polar Station in Hornsund (*red line*) and the Atlantic Water mean temperature at the transect N. **b** Regression of the both values. The 95 % confidence level is marked

Fig. 5.12 The coefficient of correlation between monthly mean air temperature in Hornsund and the AW temperature at the transect N for various time lags

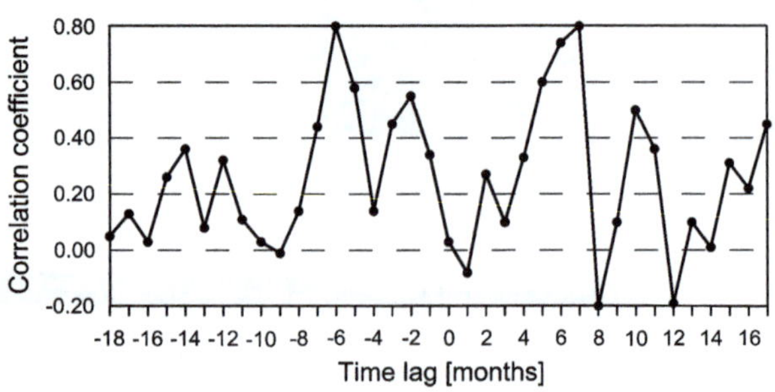

temperature influences the annual mean air temperature in Hornsund, or is it the other way round—the AW temperature is modified by the weather. The correlation between monthly averaged air temperatures in Hornsund and mean AW temperature on transect N (Fig. 5.12) was examined. The results with negative time lag represent correlations between the AW and monthly means of air temperature in Hornsund in months preceeding the hydrographic measurements, the positive time lag results represent correlations with months following the measurement. The highest correlations occur in winter months (time lag of −6 and +6 months).

The correlation coefficient between mean air temperature in three (December, January, February) winter months and the AW temperature the following summer equals **0.72**. For the AW temperature and the air temperature in 3 months of the following winter, *r* is equal to **0.80**. The results show that the interaction between the ocean and the atmosphere is bidirectional, however, the ocean impacts the air temperature stronger. Once again, the great importance of the AW for the local climate is confirmed. The result also show that the hydrographic data from the transect 'N' are particularly significant, both for oceanography and climatology.

In order to trace the change in mean temperature and salinity in different regions of section 'N', 3 parts were selected, each of 20 km width. The three parts include: the core, between the 255th and the 275th km of the transect, the central part—between the 210th and the 230th km and the western part—between the 90th and 110th km of the transect. The diagram (Fig. 5.13) clearly shows that the mean AW temperature at the transect results from a combination of mean temperatures in various parts of the transect which are influenced by different WSC branches. The western branch is colder and less saline than the core and the central WSC flow, the temperature and salinity variability is higher here. In some cases (e.g. in 2002), the values are out of phase—the increase of T and S in the core or in the central part coincides with the decrease in the western part. The variability in the western part may be related to the shift of the Arctic Front or mesoscale activity in that region. The properties in the central part are closest to the mean ones, hence the good correlation between that part and the means of all the parameters from the entire transect. The largest correlation coefficient of **0.98** was found between mean salinity and salinity in the central part, the correlation between temperatures is much lower (Table 5.1). The relations may be significant while attempting to restore the AW mean temperature from incomplete coverage of the transect.

In 2007 the values of temperature and salinity fell in all WSC branches. In the western part the temperature decrease began already in 2006.

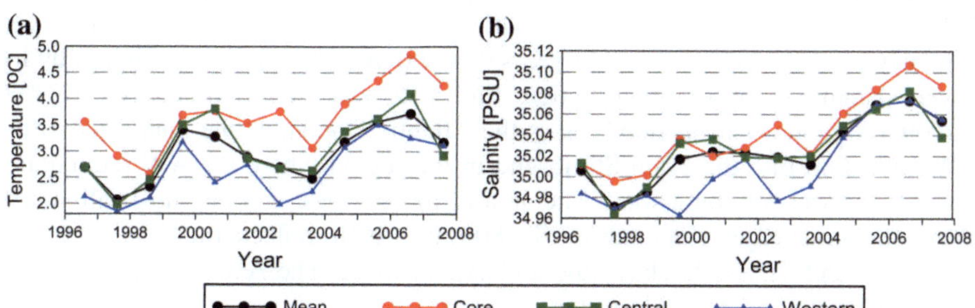

Fig. 5.13 Transect 'N'. Time series of the AW temperature (**a**) and salinity (**b**) in three areas of the transect and the mean value

Fig. 5.14 Time series of integrated heat content (TJ/m) of the AW layer in the three areas of transect 'N'

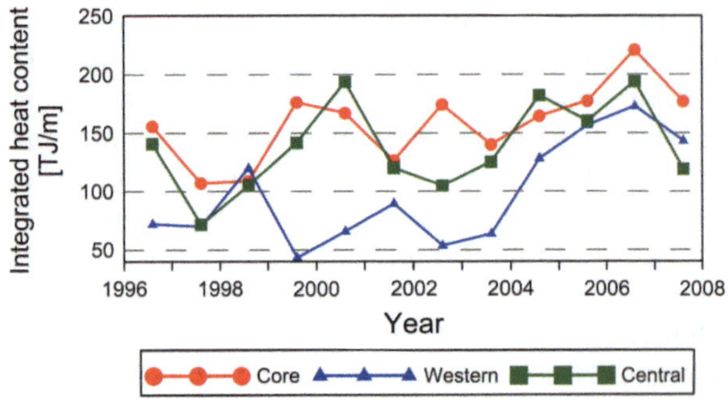

Table 5.1 The value of the correlation coefficient r between mean properties of the AW from the entire transect 'N' and properties from individual parts of the transect

Part	Temperature	Salinity	Heat content
Core	**0.90**	**0.95**	0.75
Central	**0.97**	**0.98**	**0.83**
Western	**0.89**	**0.88**	**0.79**

The diagram of integrated heat content in the AW layer in individual parts of the transect (Fig. 5.14) shows that the lowest heat content (and probably the smallest heat transport) was initially in the western part. Since 2004 the branch strengthened to an unusual extent, nearly levelling its heat content to that observed in other parts. In 2006 it must have happen due to the increased AW layer thickness, since the temperature in that part decreased (Fig. 5.13).

5.4 Transect '*EB*'

Transect 'EB' is a section located nearest to the Arctic Ocean usually measured by the IOPAS. The transect is situated in the Fram Strait and overlaps the line of current meters installed by the Alfred Wegener Institute (AWI) on moorings. The IOPAS has been performing systematic measurements since 2000, however, similarly to other sections, in cold years their extent to the west was limited by ice cover. Usually, a 160 km long section (Fig. 5.15) was measured between 01°30′E and 09°E latitude. The section crosses the Atlantic Domain at the final stage of the AW advection into the Fram Strait, and in the region where the AW recirculates towards the west or south-west. The area of the warmest and the most saline AW above the Spitsbergen slope (above the 800 m isobath) reflects the WSC core. The western branch is visible above the underwater ridge. The AW recirculates as the RAW, westward of the underwater ridge.

Despite the fact that the data are incomplete, mean temperature and salinity of the AW at the transect are presented (Fig. 5.16). Similarly to the previous sections,

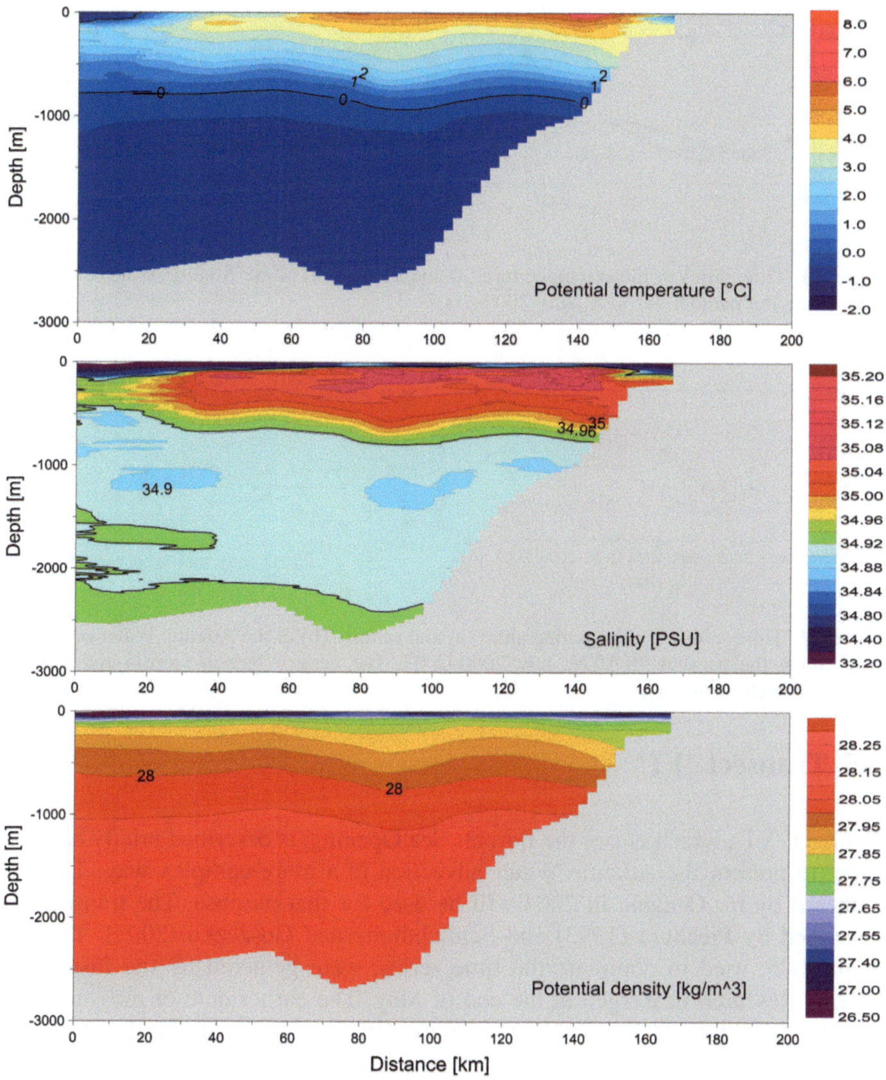

Fig. 5.15 Potential temperature, salinity and potential density on transect 'EB' along the parallel 78°50′N. Summer 2004, r/v Oceania. The 0 °C isotherm and the 34.92 isohaline are bold

including in particular 2005 and 2006, the increase in the AW temperature and salinity, and decrease of the values in 2007 is distinct here. The mean temperature and salinity in the eastern and western part of the transect were also calculated. The western part was defined as a section between 04° and 06°E latitude, and the WSC core as located between 07° and 09°E latitude. The difference in temperature and salinity of the both parts can be clearly seen (Fig. 5.17). The largest differences in temperature are in 2000 and 2003, the differences diminish after 2003.

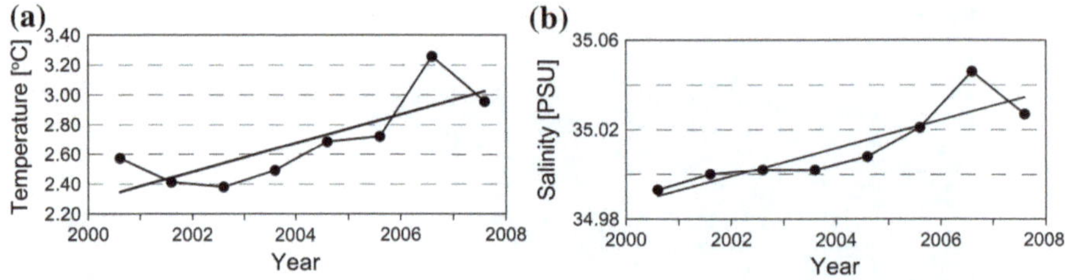

Fig. 5.16 Time series of mean temperature (**a**) and salinity (**b**) of the Atlantic Water on transect 'EB' along the parallel 78°50′N, July 2000–2007

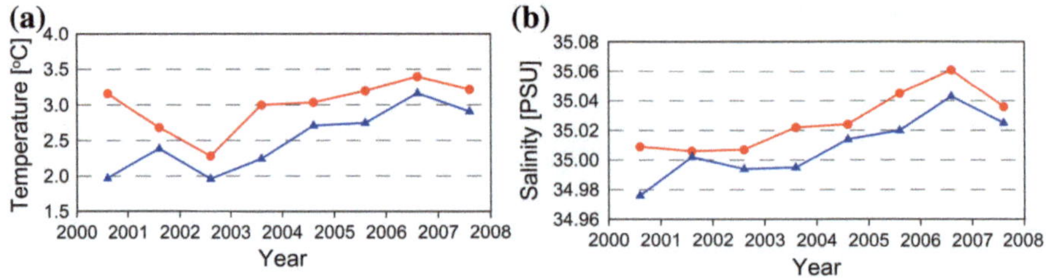

Fig. 5.17 Time series of mean temperature (**a**) and salinity (**b**) of the Atlantic Water on transect 'EB' along the parallel 78°50′N, July 2000–2007. The eastern branch (core)—*red line*, the western branch—*blue line*

5.5 Transect '*V1*'

Transect 'V1', which closes the Barents Sea Opening, is described briefly below in order to present the Atlantic Water advection in a more complex way. The data gathered by r/v Oceania in 2004 will be used for that purpose. The transect was described by Piechura (1993) and Schlichtholtz and Goszczko (2005). The data from 2005, used to complete the time series, were gathered by the Institute of Marine Research in Bergen at the end of May. The earlier date of measurements may be the reason of lower AW temperature at that transect in 2005 (Fig. 5.18).

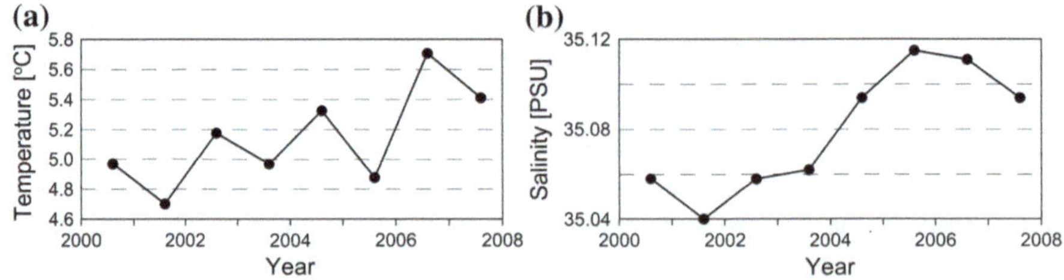

Fig. 5.18 Time series of mean temperature (**a**) and salinity (**b**) of the Atlantic Water on transect 'V1' along the meridian 20°E, July 2000–2007. The 2005 data were gathered by the IMR in May 2005

Transect 'V1' is meridional, from the northern coast of Norway—on the right of the figure, to the Bear Island. That 400 km long section crosses the main route of the Atlantic Water inflow into the Barents Sea. During its eastward advection, the AW is subject to a very strong transformation—cooling and decrease of salinity which contributes significantly to formation of deep and intermediate waters. (Rudels et al. 1994; Schauer et al. 1997, Masłowski et al. 2004). The water flows from the Barents Sea into the AO through the St. Anna trough and joins the AW inflow through the Fram Strait. The volume transport through the BSO is comparable with the volume transport through the Fram Strait. It flows through a 500 m deep trough in the shelf of the Barents Sea (Fig. 5.19). The inclination of all isolines suggests an intense baroclinic flow "towards the picture", thus, eastward. The warmest water is located on the Norwegian shelf. However, that is not only the Atlantic Water. The warm and low saline water is transported by the Norwegian Coastal Current. Low salinity of the NCC results from the advection of the fjord water mixed with water transported from the Baltic Sea (Masłowski and Walczowski 2002). The AW is the warm and saline water mass, separated by the 34.92 isohaline, which occupies the major part of the transect. Less saline Polar Water appears from the side of the Bear Island.

5.6 The 'Core' Transect

The transformation of the AW during its northward advection has influence upon its physical and chemical properties—mostly temperature and salinity. At lower latitudes the AW is warmer and more saline than in the north. That is obvious. However, the water also carries a seasonal signal, originating from the area of its ultimate formation, probably in the Subpolar Gyre. Hence the complex diagram of the AW temperature at the Svinoy Section (Orvik and Skagseth 2005). It also influences the spatial and temporal variability of the AW properties in the study area. During advection, the water also acquires its own seasonal signal; the exchange of heat with the atmosphere is enhanced in winter, it occurs with various intensity in different regions, the intensity of exchange is significantly modified by the atmospheric circulation—the wind forcing, air temperature and humidity. Therefore, the AW temperature maximum in time series from the Fram Strait occurs only in autumn, and the variability is quite complex. Assuming that it is an advective signal, it should somehow manifest upstream, south of the Fram Strait. On a hydrographic section of adequate length and resolution, the AW temperature should have a shape of a sloping sinusoid (or overlapping sinusoids) with an amplitude increasing northward (the amplification effect occurs along the northward advection). The salinity field should not obtain a distinct seasonal signal in the Nordic Seas, mean salinity on sections should decrease as the water mass flows northwards. The distance between maxima of such signal (wave length) depends on mean advection velocity. The latter value is, however, based on observations of

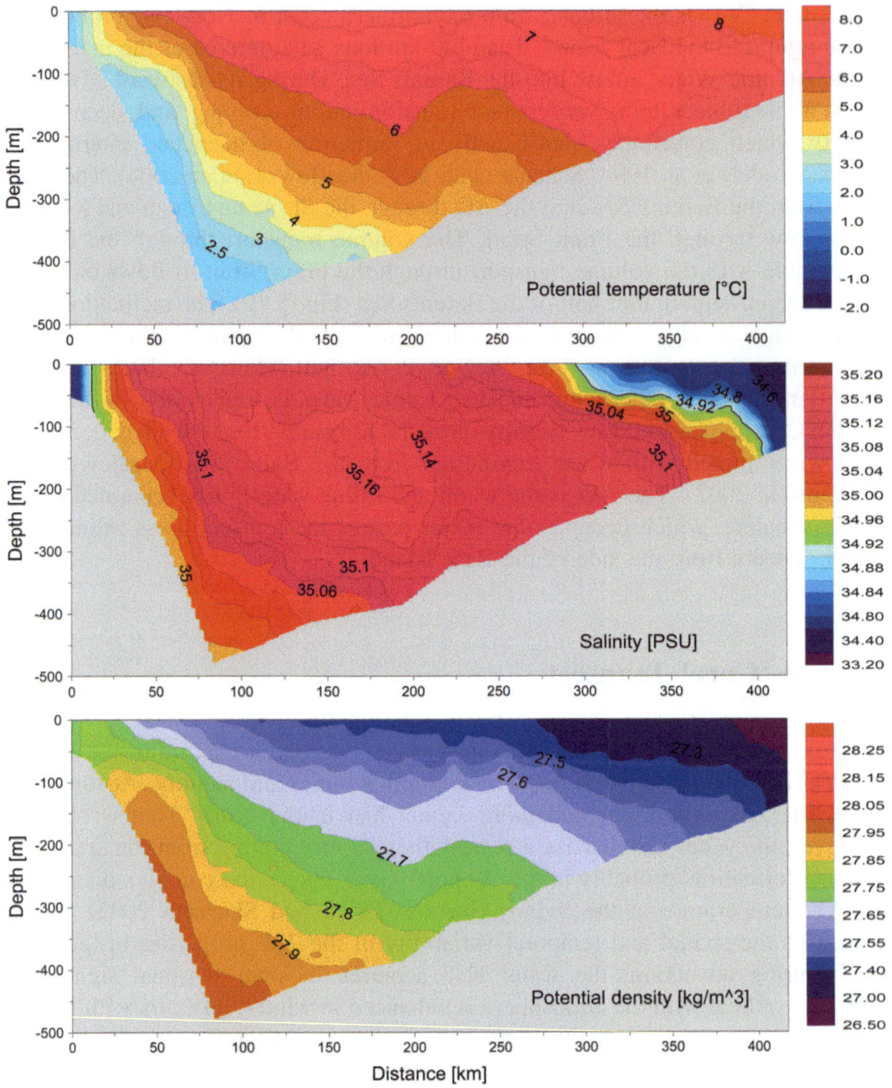

Fig. 5.19 Potential temperature, salinity and potential density on transect 'V1' between Norway and the Bear Island. Summer 2004, r/v Oceania. 34.92 isohaline is bold

specific structures propagating through the moorings with current meters (Polyakov et al. 2005) or observed at the hydrographic sections (Walczowski and Piechura 2007). The advection velocity is usually equal to a few (2–4) cm/s. In terms of the present study it is a very significant parameter, since it conditions any relationship between the observations performed in two subsequent summer seasons. With mean advection velocity of 4 cm/s, the structures observed at 70°N will pass 80°N already after a year. Furthermore, the propagation velocity may differ in various parts of the WSC. It also changes in an annual cycle.

A decision was made to investigate the variability of the AW properties in the core flowing along the continental slope. In order to perform complete sections, the coldest (2003) and the warmest (2006) years were chosen. The measurement points located in the core centre were selected from each section. These are the stations above the continental slope, at a depth of 600–1,000 m. Distributions of hydrographic properties at the sections are shown on Fig. 5.20. The sections vary significantly, both as regards temperature and thickness of the AW layer. In cold 2003 the 6 °C isotherm lies between 162 m and 36 m, and in the warm 2006 the 6 °C is located at a depth from 261 m in the southern part to 157 m up north. There is a similar slope of slope in deeper layers. Salinity curves are somewhat more complicated, however, in general, the inclination of the isohalines is similar to isotherms. Yet, it is difficult to find a seasonal signal. There may be two reasons for that—either it does not exist, or the horizontal distance between the stations is too large.

5.7 Meridional Variability

Integrated values are the most certain criteria for describing water mass properties. That enables reduction of the impact of mesoscale structures or other local processes upon the final result. It also concerns the hydrographic sections, although in that case the dynamics of the sea region has a stronger influence upon the result. Narrowing down the field of observation to smaller and smaller areas poses a risk of obtaining artificial or local results which do not reflect the general trends. Therefore, a description of meridional variability begins with integrated values—mean AW properties on sections measured by the IOPAS. This gives a certainty that objective criteria for evaluation of the processes are applied. Some of the properties were described or shown in another form (as time series) in the preceding chapter, however, thanks to the new arrangement, an integral image can be obtained. Data from the following zonal sections were used: 'H', 'K', 'N' and 'EB'. The same time period was chosen—2000–2007. The first diagrams (Fig. 5.21) show the AW temperature and salinity averaged both in space, at on the investigated sections, and in time for the entire period of study.

The average difference in temperature between the southern and northern section is equal to 1.28 °C, and in salinity 0.052, respectively. A nearly linear decrease of the AW temperature and salinity means (calculated from the entire section and all years) can be observed along the increasing latitude. For temperature it is 0.24 °C per one degree of latitude (0.21 °C/100 km), and for salinity: 0.01 per one degree of latitude (0.009 1/100 km). Standard deviation shows that the greatest variability occurs at the section 'N' along parallel 76°30'N, both for temperature and salinity. A temperature and salinity increase greater than the average occurs in the southern part of the study area. It may be a manifestation of

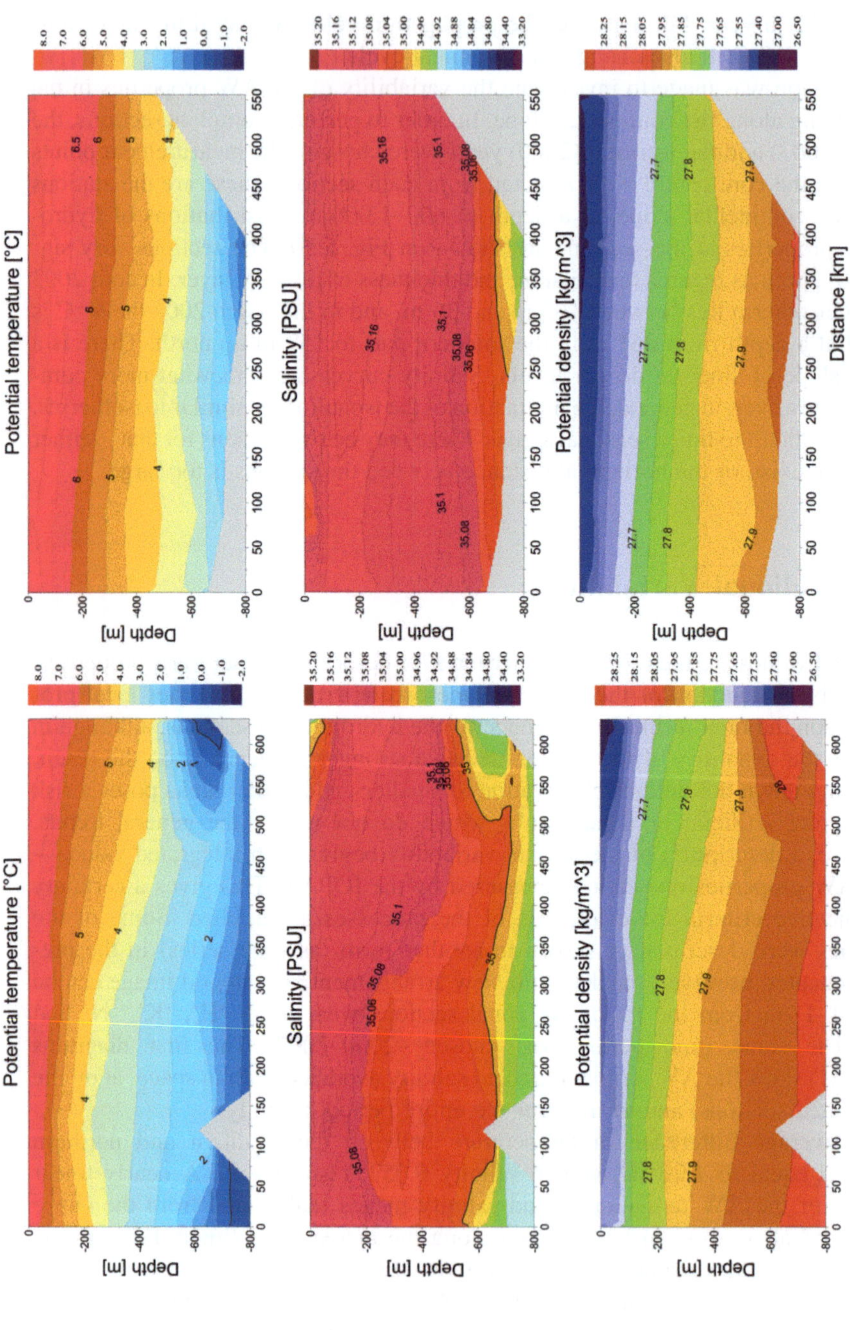

Fig. 5.20 Potential temperature (°C), salinity (PSU) and potential density (kg/m^3) on sections along the WSC core. 2003 (**a**), 2006 (**b**)

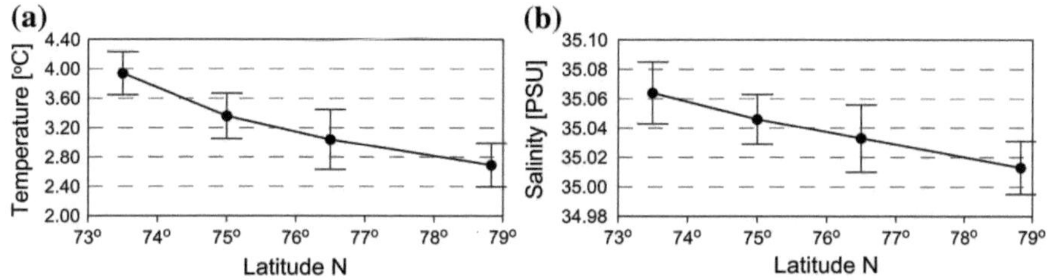

Fig. 5.21 Mean from July 2000–2007: temperature (**a**) and salinity (**b**) of the Atlantic Water on subsequent zonal sections across the WSC. Standard deviations of the presented values are marked

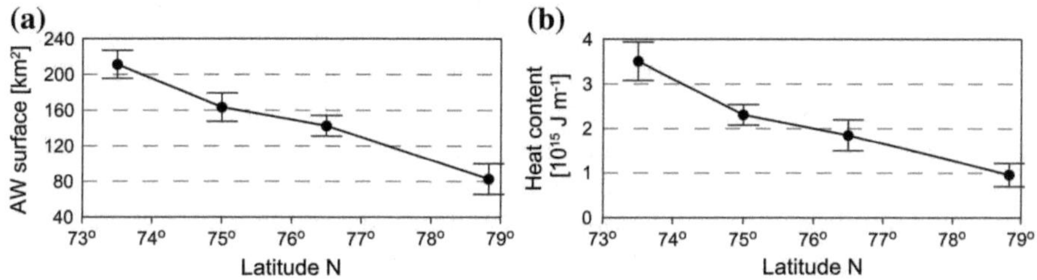

Fig. 5.22 Mean from July 2000–2007: area of section (**a**) and heat content (**b**) of the Atlantic Water on subsequent zonal sections across the WSC. Standard data deviations are marked

the effect of the AW cooling (as described further in the present paper) in the Bear Island Trough.

Also the area of the section occupied by the Atlantic Water at each transect decreases almost linearly along latitude (Fig. 5.22). The AW section area decreases almost threefold on a distance of 600 km, standard deviations, hence the interannual variability of the area, are low. Mean heat content (specified per one metre of the section thickness), decreases even more, since it is a function of section area and temperature. Similarly to temperature, of the mean heat content in the southern part is stronger (Fig. 5.22). The diagrams give the idea of the scale of change in the northward heat and volume flux and, most of all, of the scale of exchanges that occur on the boundaries of the Atlantic Domain, the Barents Sea and the Greenland Sea. It is distinct that the processes of exchange have not been properly appreciated so far.

The diagrams showing temperature change as a function of latitude in subsequent years (Fig. 5.23) confirm that the decrease of temperature towards the north is almost linear, and the angles of inclination of plot lines for individual years are very much similar to the average value. The largest change of inclination of plot lines occurs at the transect N. It is well visible particularly in 2003, when temperature is extremely low.

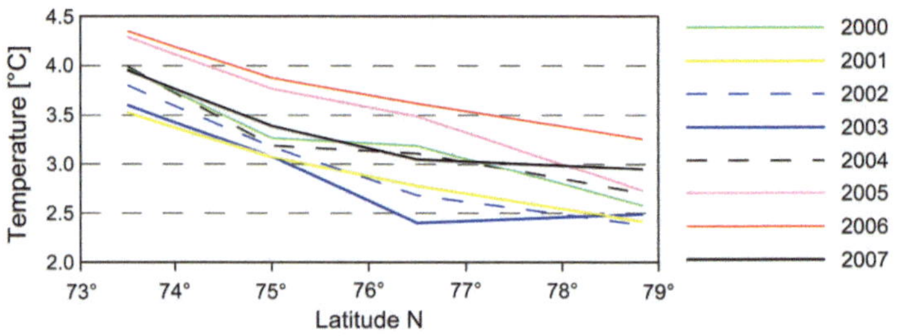

Fig. 5.23 Change in the AW mean temperature in function of latitude, calculated on transects H, K, N, EB in subsequent years

Figure 5.23 shows that the years 2001–2003 were cold along the entire length of the study area, 2000 and 2004 were years of temperature close to average and temperatures in 2005 and 2006 were much above the average. In 2005 the warming could be seen at lower latitudes; in summer 2006 temperature in the Fram Strait reached extremely high value never recorded before. That was attributed mostly to the propagation of anticyclonic eddies of unprecedented size, carrying great temperature and salinity anomalies (Walczowski and Piechura 2006, 2007). In 2007 cooling proceeded gradually from the south.

5.8 Correlations Between the Properties of the Atlantic Water Studied at the Transects

In further part a decision was made to investigate whether there are any relations between the properties of the AW at the transects. The figure presenting time series of temperature and salinity at individual transects (Fig. 5.24) shows the similarity between the time series. The values of mean temperature and salinity at all transects rose from 2005 to 2006 and decreased afterwards. The greatest deviation from the mean was revealed by temperature and salinity at the transect N in 2003 when both values were significantly lower as compared to other sections.

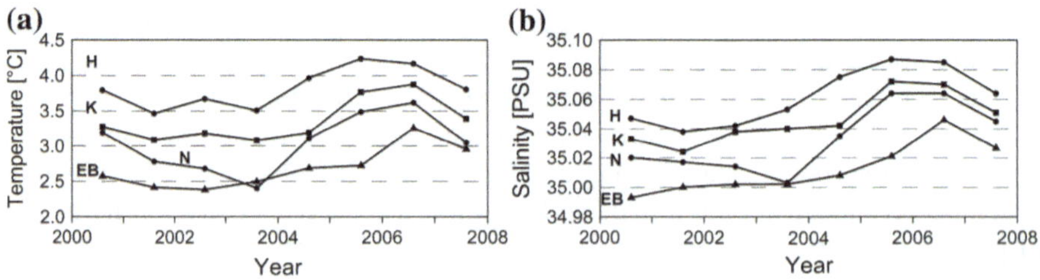

Fig. 5.24 Time series of mean temperature (**a**) and salinity (**b**) of the AW at the transects 'H', 'K', 'N' and 'EB'

Table 5.2 The correlation coefficients r between the AW mean temperature at pairs of transect

Transect	K (2000–2007)	N (2000–2007)	EB (2000–2007)
H (2000–2007)	**0.81**	**0.93**	0.51
K (2000–2007)		**0.86**	0.64
N (2000–2007)			**0.74**

Table 5.3 The correlation coefficients r between the AW mean salinity at pairs of transect

Transect	K (2000–2007)	N (2000–2007)	EB (2000–2007)
H (2000–2007)	**0.77**	**0.77**	0.37
K (2000–2007)		**0.73**	0.56
N (2000–2007)			0.63

Table 5.4 The correlation coefficients r between mean temperature and mean salinity of the AW at transects

Transect	H (2000–2007)	K (2000–2007)	N (2000–2007)	EB (2000–2007)
H (2000–2007)	**0.87**	0.69	**0.78**	0.43
K (2000–2007)		**0.91**	0.68	0.37
N (2000–2007)			**0.96**	0.58
EB (2000–2007)				**0.80**

Correlations and p-tests between the T and S time series of all the transect pairs were calculated for detrended data. The results are given in Tables 5.2, 5.3 and 5.4. Correlations on significance level below 0.05 are marked bold, below 0.01—bold and underlined.

Correlations on significance level below 0.05 were obtained for the temperature–temperature and salinity–salinity pairs for transect pairs nearest to each other: H–K, H–N, K–N; and for the pair N–EB—only for the temperature–temperature pair. It is interesting that the correlation between temperatures on transects H–N is higher than the correlation between transects H–K which are closer to each other. That applies also to the correlation between salinity on transect H and temperature on transect N. That may be a manifestation of the sought seasonal signal carried by the AW, since for salinity on transects H–N the correlation is only slightly higher than for transects H–K. Unfortunately, there are not enough data to state the above with full certainty. Besides, if the same signal phase was observed at the transects H and N (e.g. some "recorded" subsequent summer signals, as suggested by Fig. 5.21), the propagation velocity would be very low. To travel 3 degrees of latitude per year (from 73°30′ to 76°30′), the average propagation velocity should be around 1 cm/s. The shift of correlation of temperature in relation to salinity cannot be seen—T and S correlate very well on each transect (Table 5.4).

Calculating correlations (and autocorrelations for pairs of the same transects) with 1 year time lag also gives interesting results (Tables 5.5, 5.6).

Table 5.5 The correlation coefficients r of the AW mean temperature at individual transects with 1 year time lag

Transect	H (2001–2007)	K (2001–2007)	N (2001–2007)	EB (2001–2007)
H (2000–2006)	−0.02	0.50	0.32	0.53
K (2000–2006)		0.05	0.02	0.54
N (2000–2006)			0.22	0.40
EB (2000–2006)				−0.11

Table 5.6 The correlation coefficients r of the AW mean salinity at individual transects with 1 year time lag

Transect	H (2001–2007)	K (2001–2007)	N (2001–2007)	EB (2001–2007)
H (2000–2006)	0.23	0.38	0.70	0.67
K (2000–2006)		−0.16	0.09	0.56
N (2000–2006)			0.24	0.65
EB (2000–2006)				−0.16

Table 5.7 The correlation coefficients r of mean salinity and temperature (columns) of the AW shifted by 1 year

Transect	H (2001–2007)	K (2001–2007)	N (2001–2007)	EB (2001–2007)
H (2000–2006)	0.42	0.70	0.71	0.67
K (2000–2006)		0.17	0.22	**0.79**
N (2000–2006)			0.23	0.50
EB (2000–2006)				−0.06

The value of the autocorrelation function for 1 year time lag is very low—it means that based on 1 year's study of a single transect we cannot conclude on the AW properties in the following or the preceding year. However, salinity at the transect N can be predicted with high probability based on the salinity data from the transect H in the preceding year. It is quite a general rule that in the case of one-year time lag between the transects situated in the north in relation to those in the south, the more distant transects are correlated better. It indirectly confirms the thesis that time needed for propagation of the signal from the south to the north approximates to 1 year. The highest r (although on the significance level above 0.05) is for the temperature of H–EN and K–EN pairs, and for the salinity of H–N, H–EB and N–EN pairs. It suggests that mean salinity and temperature of the AW in the Fram Strait react with 1 year lag in relation to processes which occur south of transect EB. The strongest correlations are obtained between salinity and temperature in the following year. In this case the correlation coefficient increases significantly as the distance between the sections increases (Table 5.7).

When the study area was divided into two parts, the temperature in the northern part also reacted with 1 year lag in relation to salinity in the southern part (Chap. 4).

5.8.1 Correlations in the Eastern WSC Branch

For the eastern branch (core) of the WSC, mean AW properties were calculated in the WSC core, in a similar way as for the entire transects. A section of 50 km width has been selected from each transect to cover the area of the highest temperature and salinity. This selection of the data subset alone causes the data to be less representative than the data from the entire transect, since the position of the shifts in relation to the shelf and its width vary in years. However, a decision was made not to distinguish any regions separately for each year, since that may lead to subjective selection of data. Temperature and salinity time series from the WSC core (Fig. 5.25) at individual transects differ significantly. Transect H, which lies farthest to the south features the lowest variability, although it is situated in the range of impact of the Barents Sea. The difference between the minimum in 2001 (4.15 °C) and the maximum in 2006 (4.88 °C) equals 0.73 °C, and standard deviation: 0.24 °C. For the purpose of comparison: at the transect N temperature reaches minimum in 2003 (3.05 °C), maximum in 2006 (4.10 °C) and its standard deviation is 0.45 °C. It indicates that variability of properties of the AW flowing from the NwASC is relatively low. However, there is a significant change in properties of the AW as it continues northwards. Temperature and salinity decreases, differences between the properties in different years grow. It is sometimes referred to as "signal amplification".

In most cases, the correlation coefficients for the WSC core temperatures are lower than the ones for mean temperatures from the entire transect (Table 5.8). A significant correlation was obtained for the pair N–EB. The correlation coefficients

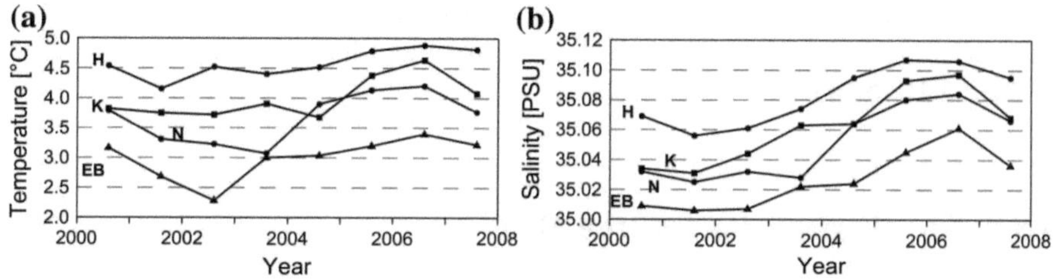

Fig. 5.25 Time series of mean temperature (**a**) and salinity (**b**) of the AW in the WSC core on transects H, K, N and EB

Table 5.8 The correlation coefficients r between the AW mean temperature of transect pairs in the WSC core

Transect	K (2000–2007)	N (2000–2007)	EB (2000–2007)
H (2000–2007)	0.42	0.62	0.57
K (2000–2007)		0.50	0.60
N (2000–2007)			**0.77**

Table 5.9 The correlation coefficients r between the AW mean salinity of transect pairs in the WSC core

Transect	K (2000–2007)	N (2000–2007)	EB (2000–2007)
H (2000–2007)	**0.75**	**0.89**	0.59
K (2000–2007)		0.59	**0.71**
N (2000–2007)			**0.73**

Table 5.10 The correlation coefficients r of the AW mean temperature on individual transects in the WSC core with 1 year time lag

Transect	H (2001–2007)	K (2001–2007)	N (2001–2007)	EB (2001–2007)
H (2000–2006)	−0.50	0.39	−0.12	0.42
K (2000–2006)		−0.15	−0.12	0.08
N (2000–2006)			0.24	0.32
EB (2000–2006)				0.11

Table 5.11 The correlation coefficients r of the AW mean salinity on individual transects in the WSC core with 1 year time lag

Transect	H (2001–2007)	K (2001–2007)	N (2001–2007)	EB (2001–2007)
H (2000–2006)	0.24	0.11	0.52	0.57
K (2000–2006)		−0.12	0.35	0.29
N (2000–2006)			0.15	0.49
EB (2000–2006)				−0.19

for salinity are comparable, and for the pair H–N the correlation coefficient reaches **0.89** (Table 5.9).

Autocorrelations and correlations between the same properties of the AW at various transects calculated with 1 year time lag have usually lower values (Tables 5.10, 5.11).

5.8.2 The Western Part of the WSC

The same calculations as for the eastern part were performed for the western part of the WSC (Fig. 5.26), (Tables 5.12, 5.13).

Similarly to the entire transects, also in the western part the concurrent correlations between temperatures for pairs H–K, H–N and K–N are the highest. In the case of salinity only the K–N pair indicates a correlation at the significance level below 0.05, while for the correlation with a time lag, only the salinity of the H–N pair exceeds that level. However, it is typical that in all the cases with time lag, correlations for transect EB are larger in the western part than in the core. For salinity of the N–EB pair, the coefficient of correlation reaches 0.72 (Table 5.14).

All the operations yield interesting, yet incomplete, results.

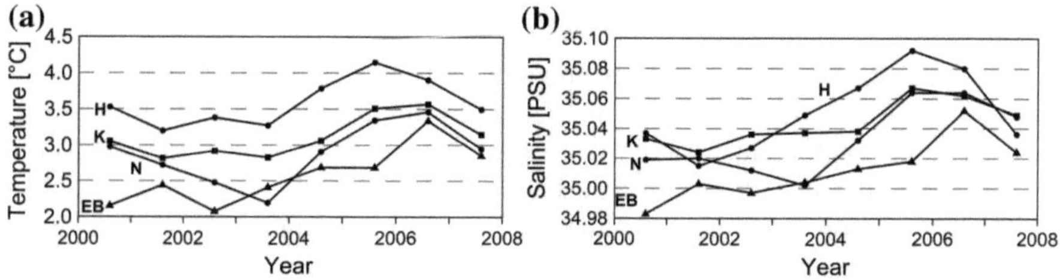

Fig. 5.26 Time series of mean temperature (**a**) and salinity (**b**) of the AW west of the WSC core on transects H, K, N and EB

Table 5.12 The correlation coefficients r of the AW mean temperature for transect pairs west of the WSC core

Transect	K (2000–2007)	N (2000–2007)	EB (2000–2007)
H (2000–2007)	**0.84**	**0.84**	0.25
K (2000–2007)		**0.90**	0.45
N (2000–2007)			0.60

Table 5.13 The correlation coefficients r of the AW mean salinity for transect pairs west of the WSC core

Transect	K (2000–2007)	N (2000–2007)	EB (2000–2007)
H (2000–2007)	0.69	0.67	0.12
K (2000–2007)		**0.74**	0.35
N (2000–2007)			0.53

Table 5.14 The correlation coefficients r between the AW mean temperature and the AW mean temperature with 1 year time lag for transects situated west of the WSC core

Transect	H (2001–2007)	K (2001–2007)	N (2001–2007)	EB (2001–2007)
H (2000–2006)	0.15	0.57	0.59	0.66
K (2000–2006)		0.03	0.09	0.55
N (2000–2006)			0.26	0.40
EB (2000–2006)				−0.24

Correlations without a time lag.

1. Most of the correlation coefficients between the investigated mean values from the entire transects are higher than the coefficients between the analogical values from “partial” transects.
2. The highest correlations, both for mean temperature and salinity were found between transect H and N;
3. In the western part there are also high correlations between transects K and N. They do not exist in the WSC core.
4. In the WSC core, there are large correlations between transects N and EB.

Correlations calculated with time lag

1. The value of the autocorrelation function is very low in all cases;
2. Both for mean salinity and temperature 1-year time lag did not increase the correlation coefficients. They are comparable or lower than respective coefficients calculated without the time lag;
3. The correlation coefficients in the western part are higher than in the entire transects or in the eastern part;
4. There are stronger correlations between salinity at the southern transects and temperature at the northern ones in the following year (Table 5.7).

The results give basis to draw a few more general conclusions concerning the nature of spatial and temporal variability of the AW properties in the given area.

1. Based on data from one year one cannot predict mean values of the AW in the following year on the same transect, since temperature and salinity in the following year are not correlated with temperature in the current year.
2. Based at temperature at the transects one can predict with some probability the temperature and salinity in the downstream region. In this particular case, one can predict the AW mean temperature and salinity at the transect 'N' in the same year with high probability, for instance from transect H data. These values are well correlated (Fig. 5.27).
3. Despite the long periods of variability and advective nature of "the ocean climate" in the study area, the possibility of predicting temperature at northern transects in the following year based on the temperature data from the southern transects is limited. The same is with salinity. 1-year time lag was observed for some pairs of transects, however, the correlation coefficients were at the significance level above 0.05. Correlations in the WSC core were even lower. They increased slightly in the western part. In that case the correlation between the values of mean salinity at transects H and N in the following year was 0.77 (Table 5.15).

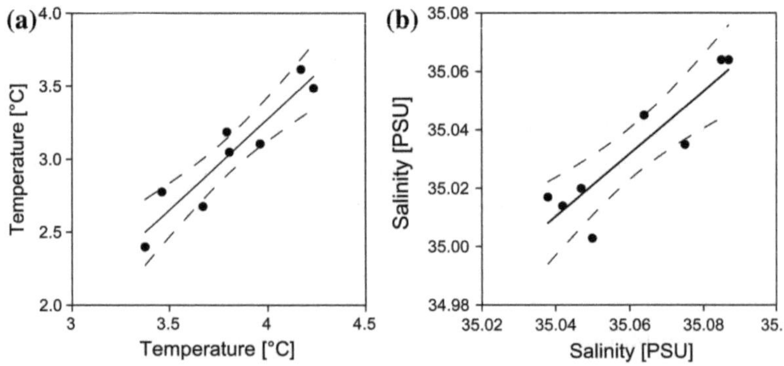

Fig. 5.27 Regression of mean temperature (**a**) and salinity (**b**) of the AW of the H–N transects pair. The data are not detrended. 95 % confidence intervals are marked

Table 5.15 The correlation coefficients r between the AW mean salinity and the AW mean salinity with 1 year time lag for transects situated west of the WSC core

Transect	H (2001–2007)	K (2001–2007)	N (2001–2007)	EB (2001–2007)
H (2000–2006)	0.25	0.46	**0.77**	0.66
K (2000–2006)		−0.17	0.02	0.64
N (2000–2006)			0.27	0.72
EB (2000–2006)				−0.14

$$T_N = 1.2479 * T_H - 1.7177 \quad r = 0.937$$
$$S_N = 1.0717 * S_H - 2.5420 \quad r = 0.814$$

It is worth emphasizing again that the time series we have at our disposal are short, hence, not statistically robust. However, they are the only series from such a large area. Time series are extended each year and the measurement will be the best way to verify the presented relationships. There are a lot of questions. It is not completely clear why the correlation is generally higher (without any time lag) for more distant transects at (K–N) than closer to each other (H–N). Is it a seasonal signal occurring at an average distance of 2.5-3 degrees of latitude (275-330 km). That is a too short distance taking into account the described AW advection velocity. On the other hand, according to observations, the temperature isoline shifted northwards with average annual velocity of 1 cm/s (Walczowski and Piechura 2007), which is in conformity with the above distances. However, shifting of isotherms does not necessarily mean that there is a movement of water; during advection some heat is released to the atmosphere, thus the shifting of isotherms is slower than the movement of water particles. The issue will be addressed in Chap. 8.

Obtaining correlations between the data with 1-year time lag is a very valuable result. According to the statistics, we are capable of maintaining some continuity of observations by performing measurements only once per year. Water masses observed one year in the south or in the central part of the study area may be observed the following year in the northern part, in the Fram Strait area. It concerns in particular the processes occurring in the western part of the WSC where the average advection velocity is lower. There is a higher correlation between the values of salinity, since that property is more conservative. For instance, transect 'EB' shows the unique role of processes in Fram Strait and in the entire region west of Spitsbergen. The highest correlations with other sections were obtained at that transect for 1-year time lag in particular. It confirms the thesis that the upstream region (south of the 78°N parallel) may be a sort of a buffer for the AW.

5.9 Summary

This chapter focuses on presentation and analysis of variability of the AW properties on spatial scales smaller than the one applied in Chap. 4. Therefore, mainly the variability at transects has been presented. They mostly include zonal sections, from the most southern ones to the northern ones. Two meridional transects have also been used: transect V1 between Norway and the Bear Island, and a virtual Core transect, developed for the purpose of the present work.

Similarly to mean values from the entire study area, a distinct warming in the 2004–2006 period is visible, a phase lag can be also seen in the propagating signal. In order to determine the structure of warming, time series of the AW properties at the same transects but in different parts (western, central and eastern) were studied. In the case of the transect N time series, the Hovmoeller diagram was used.

The data from transects were used for more detailed analysis of the transformation which the Atlantic Water undergoes along its northward advection. A linear decrease of the AW mean properties, such as temperature, salinity, heat content or the cross-section area occupied by the AW, was found.

Relationships and evidence to confirm the advection-driven propagation of signal was sought with use of a series of correlations between the AW properties at transects. The most valuable result is the correlation between water properties on sections H and K (the southern study area) and the AW properties in the northern study area the following year, increasing on its way northwards. It confirms once more that the average time of signal propagations through that region is approximately 1 year. That applies mostly to the western branch; in the WSC core the signal propagation speed is higher.

References

Druet C (1994) *Dynamika Stratyfikowanego Oceanu*, PWN, p 225

Maslowski W, Walczowski W (2002) Circulation of the Baltic Sea and its connection to the pan-Arctic region—a large scale and high resolution modeling approach. Boreal Environ Res 74:319–325

Maslowski W, Marble D, Walczowski W, Schauer U, Clement JL, Semtner AJ (2004) On climatological mass, heat, and salt transports through the Barents Sea and Fram Strait from a pan-Arctic coupled ice-ocean model simulation. J Geophys Res 109:C03032. doi:10.1029/2001JC001039

Orvik KA, Skagseth Ø (2005) Heat flux variations in the eastern Norwegian Atlantic Current toward the Arctic from moored instruments, 1995–2005. Geophys Res Lett 32:L14610. doi:10.1029/2005GL023487

Piechura J (1993) Hydrological aspects of the Norwegian-Barents confluence zone. Studia i Materialy Oceanologiczne, Polar Marine Research 2(65):197–222

Polyakov IV et al (2005) One more step toward a warmer Arctic. Geophys Res Lett 32:L17605. doi:10.1029/2005GL023740

Rudels B, Jones EP, Anderson LG, Kattner G (1994) The Polar Oceans and their Role in Shaping the Global Environment. Geophysical Monograph Series. vol 85, American Geophysical Union, Washington D.C., pp 33–46

Schauer U, Muench RD, Rudels B, Timokhov L (1997) The impact of eastern Arctic Shelf waters on the Nansen Basin intermediate layers, J Geophys Res 102, 3371–3382

Schlichtholz P, Goszczko I (2005) Was the Atlantic water temperature in the West Spitsbergen Current predictible in the 1990s?. Geophys Res Lett 32:L04610 doi: 10.1029/2004GL021724

Walczowski W, Piechura J (2006) New evidence of warming propagating toward the Arctic Ocean. Geophys Res Lett 33:L12601. doi:10.1029/2006GL025872

Walczowski W, Piechura J (2007) Pathways of the Greenland sea warming. Geophys Res Lett 34:L10608. doi:10.1029/2007GL029974

Chapter 6
Dynamics of the West Spitsbergen Current

6.1 Data and Methods

Despite a significant progress in technology, measuring ocean currents in open ocean from the ship remains difficult and the results obtained have large errors. It is affected by several factors. The first of them is the point of reference. Measurements are performed from an observation platform which moves in all planes, and vertical accelerations are sometimes significant. The ship's drift velocity is usually higher than the measured speed of sea current; the ship is concurrently adrift by the wind and the surface current. Thus, the best method of measuring sea currents is based on moorings, equipped with current meters distributed along the vertical line and gathering time series at many levels. However, that method is expensive, inconvenient, unreliable and the results are not always accurate.

The second reason which makes once measured currents not representative for an average flow is a significant spatial and temporal variability of the studied phenomenon. Our experience (Walczowski et al. 2005) shows that the direction of the West Spitsbergen Current may change by 180° on a time scale of a few days. Sea currents may flow in opposite directions in distances of several kilometres from each other. Sea currents vary also vertically—both as to their direction and flow velocity. Deep currents are usually barotropic—generated by sea surface tilt, vertically homogenous. At the surface they overlap with baroclinic flows—generated by the horizontal inhomogeneity of the water density and Ekman currents—induced by the friction of wind on water surface. The flows combine with tidal currents.

Significant progress has been made in the measuring technology, including introduction of the global satellite positioning system (GPS) to aid accurate determination of a ship's position, dynamic ship positioning, Doppler current profilers which measure current velocity in a lot of vertical cells, and fast computers capable of processing vast amounts of data. Yet, we cannot modify the nature of sea currents, thus, currents vectors obtained from the in situ measurements, presented in the paper, should be treated as momentary values.

W. Walczowski, *Atlantic Water in the Nordic Seas*,
GeoPlanet: Earth and Planetary Sciences, DOI: 10.1007/978-3-319-01279-7_6,

Several methods were applied when determining velocities of sea currents. During the AREX cruises continuous measurements were performed with the vessel-mounted acoustic Doppler current profiler (VM-ADCP). The device measures the direction and velocity of sea currents in a water column to a depth of 300 m. Current vectors are calculated in cells of specific thickness (usually 10 or 20 m). Introducing the ADCP measurements revolutionized the way currents are studied; now, we obtain huge amount of data, however, the accuracy is still insufficient and depends on many factors, particularly sea surface waving. Another device which uses the Doppler effect in sea measurements is a lowered acoustic Doppler current profiler (LADCP). The IOPAS has been using it since 2003 to obtain interesting and sometimes ground-breaking results. The device works according to the same principles as the hull-mounted ADCP but it is lowered, usually attached to the bathymetric rosette and the CTD sampler. It enables measuring full profile of sea currents, from the surface to the bottom. The measurements are performed at a constant frequency (usually 1 Hz) in cells of a few metres, in a layer of 200 m thickness. Thus, each measuring 'ping' provides large quantity of data from various levels, recorded in the device memory. When the rosette is lowered and raised at 1 m/s, profiling down to a depth of 2,000 m takes more than one hour, which allows to collect enough data for statistical processing. Generally speaking, the processing features separating the absolute movement—sea currents—from the relative movement of the instrument. The values overlap in the recorded set of results. 1-s averaged GPS data, providing information about the ship's position and movement, are used to define the absolute movement. However, it does not provide any information about the movement of the set of instruments lowered on the cable. The profiler is equipped with sensors which record its pitch and roll, however, under strong wind the ship's drift is fast and it is often the case that 3000 m of cable must be paid out to reach a depth of 2,000 m. The reconstructed movement of the profiler is then subtracted from the measured data. Software developed by the author of the paper and the LDEO package, published by Visbeck and Klahmann (http://www.ladcp/ldeo.columbia.edu/ladcp/) according to Fisher and Visbeck's methods, were used to process the LADCP data. Error of current profiles largely depends on the conditions, particularly sea state during profiling. Yet, it is the only method which enables measuring full profile of sea currents from ship.

Finally, tidal currents are from the processed currents measurements by subtracting the calculated vector of tidal currents from the measured values.

Due to the above reasons, each profile of sea currents taken from aboard the ship should be, with some exceptions, treated as instantaneous and local. That is why other methods, yielding more general and integrated results, are applied.

General direction of advection may be deduced from the shape of isolines of temperature, salinity or other physical properties. Mean flow usually follows the isolines of those values. Average climatic value of the signal propagation velocity can be also estimated from the time series of the measured values. If the observed structures exist for time span longer than 1 year, it is also possible to define mean velocity of their movement from the measurements taken year after year.

However, a reservation should be made that they are average climate values which are much lower than the measured velocities—e.g. for an eddy, the linear velocity of the current on its edges may reach tens of cm/s, and the propagation velocity of the entire structure—1 cm/s.

The hydrographic data and information contained therein regarding water masses is also used for calculation of velocities of baroclinic geostrophic currents with use of geopotential heights. Unlike the results of in situ measurements, the calculations yield averaged, steady flow, representative for the climate values. It is the essence of the method—it is used for calculating the values of flows from distribution of water masses and their horizontal inhomogeneity which do not change on short time scales. The geostrophic calculations are based on the assumption of steady flow generated by the horizontal gradient of hydrostatic pressure and the Coriolis force. For the horizontal movement, equilibrium is assumed between those forces, the vertical movement is forced by gravity:

$$\frac{\partial p}{\partial x} = \rho f v; \quad \frac{\partial p}{\partial y} = -\rho f v; \quad \frac{\partial p}{\partial z} = -\rho g \tag{6.1}$$

where ρ is the density of sea water, f is the Coriolis parameters $f = 2\,\Omega \sin \varphi$, Ω is the angular speed of Earth's rotation, φ—latitude.

In the oceanographic practice, the hydrostatic pressure gradients on isobaric surfaces are calculated from the geopotential height anomaly $\Delta\Phi$, expressed in the geopotential metres (m^2/s^2 = J/kg), while applying the specific volume anomaly δ (m^3/kg),

$$\Delta\Phi_A = \int_{p_1}^{p_2} \delta dp \tag{6.2}$$

The perpendicular component V of the relative velocity between two hydrographic stations situated at a distance L may be calculated from the following formula:

$$V = \frac{(\Delta\Phi_B - \Delta\Phi_A)}{2\,\Omega L \sin \phi} \tag{6.3}$$

The weakness of the method is that it provides only baroclinic currents, since only the geopotential height resulting from the anomaly of density (or specific volume) may be calculated from the hydrographic data. In practice, it causes the necessity of selecting a no-motion level (NML), where no flow is assumed a priori. Nevertheless, the method is suitable for climate analyses, particularly when their purpose is to investigate the thermohaline circulation. Even if this method does not provide an accurate value of flow or transport, for the analysis of time series it provides a relative information—for instance, that the baroclinic flux grew two-fold. The baroclinic currents also show mean directions of water mass advection well. The method is used for horizontal distributions and vertical profiles. In the case of the profiles only the current component perpendicular to the section is obtained.

Determining the NML, for which we a priori assume that the currents are zero, is a separate problem. A NML of 1,000 m was used to calculate the presented horizontal distributions. That is justified by the fact that the vertical extent of the studied Atlantic Water does not exceed that depth. Similarly to the temperature or salinity field, for each measuring station and at each level, first, specific volume anomalies δ are determined, and then the geopotential anomalies (GPA) $\Delta\Phi$ are calculated by integration of the specific volume anomaly from the NML. The values of $\Delta\Phi$ are interpolated onto a regular grid with use of the same objective interpolation methods as temperature or salinity. For stations shallower than the NML, missing δ data missing to the NML level were filled by the values interpolated from the nearest deep stations. The method produces zero values at the bottom.

Measurements of currents in the Fram Strait with use of moored current meters indicate that there is a significant barotropic component here (Hanzlick 1983; Fahrbach et al. 2001) implying non-existent NML. That is why the geostrophic method is combined with in situ measurements by determining the barotropic reference velocity from the ADCP, LADPC or the moored current meters, and combining them with baroclinic calculations (Osiński et al. 2003) and inversion methods (Rudels 1987; Schlichtholz and Houssais 1999).

Due to the used geographic projection, the calculated U and V components of the current vector are not aligned with the eastern and northern components but the components along the axis X and Y of the Cartesian coordinate system. The system begins in the point of the following coordinates: $\varphi = 69°$ N, $\lambda = 000°$. Since the WSC axis does not align with the direction of coordinates in the chosen projection, the velocity vector components were calculated for the coordinate system rotated 30° counter-clockwise, to align with the WSC axis, in addition to U and V components. They were called U_{30} and V_{30}. On the meridian 030°E the U_{30} component is directed to the west, V_{30}—to the north, and the farther from the meridian, the more the components will deviate from the eastward and northward direction. The angle of rotation was selected to ensure that the coefficient of correlation between the U_{30} and V_{30} components is minimum. Near the Fram Strait, the WSC direction conforms to the V_{30} component, while the direction of the flow to the Barents Sea through the Barents Sea Opening is aligned with the U_{30} component.

The flow intensity and water advection routes are well reflected by the kinetic energy of the flux:

$$KE = \frac{V^2}{2}\left[\frac{\mathrm{cm}^2}{\mathrm{s}^2}\right] \tag{6.4}$$

The presented value is not in fact energy (it does not even have an adequate magnitude—it is energy per mass unit), yet it is widely used in that form due to the convenience in use. As a scalar value, it is easier to present and more suitable for averaging.

6.2 Baroclinic Currents

The baroclinic calculations provide steady currents, generated by the difference in water density, and fluxes and transport are close to the climate values. Hence, these results are presented first.

July 2000–2007 mean of the geopotential anomaly and geostrophic currents (Fig. 6.1) indicates the main directions of the AW propagation. The GPA may be treated here as a stream function. The largest dynamic heights occur in the south-eastern part of the study area, near the coastline of Norway. It is related to the warm and relatively low saline Norwegian Coastal Current. Low salinity and high temperature yield the low density and, consequently, large dynamic heights or geopotential anomalies in that region. The inflow of the NCC into the Barents Sea is clearly visible. Along with the NCC, the Atlantic Water carried by the NwASC flows. The second area of intense AW inflow is situated in the south-western part of the study area, above the underwater Mohn Ridge. A region of weak and variable fluxes is visible between the both branches. Between the latitudes of 72°N and 74°N the lines of geopotential anomalies are aligned, similarly to temperature and salinity isolines, almost along the parallels. The arrangement of the stream-lines changes into meridional north of the parallel 74°N. The effects of such a configuration can be seen in the vectors of the baroclinic currents—an eastward

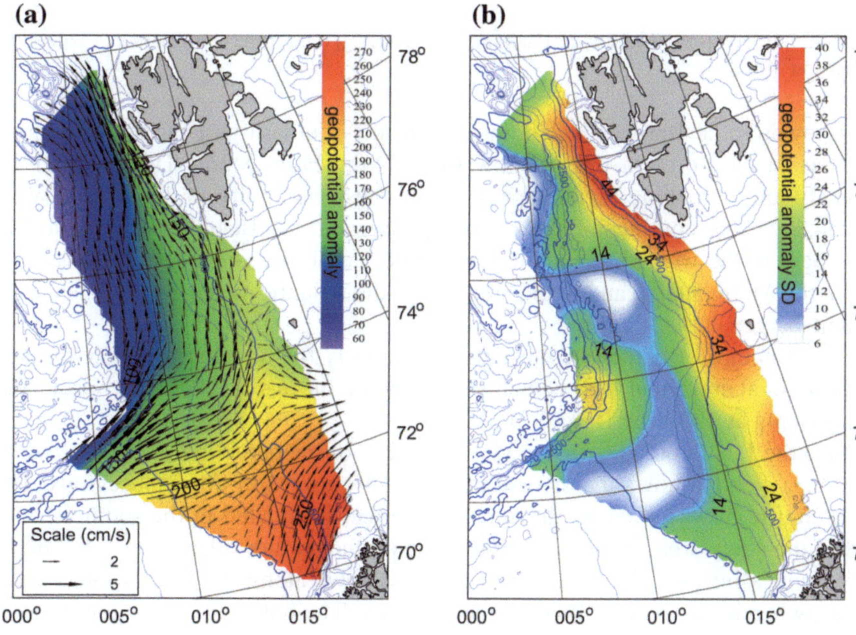

Fig. 6.1 July 2000–2007 mean of the geopotential anomaly (10^{-6} m^2/s^2) at 100 dbar level depicted by the colour scale and mean baroclinic **a** calculated with the NML at 1,000 dbar **b** Standard deviation of the geopotential anomaly

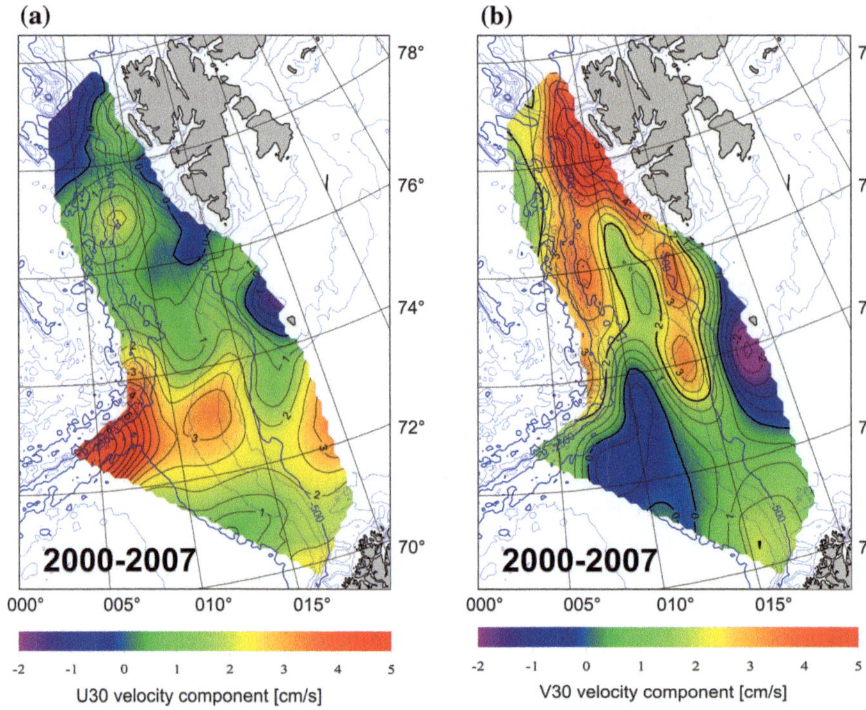

Fig. 6.2 July 2000–2007 means of U_{30} (cm/s) (**a**) and V_{30} (cm/s) (**b**) components of baroclinic currents at 100 dbar

flow, towards the Barents Sea, dominates in the southern part of the study area, whereas in the northern one—a flow towards the Fram Strait. The geopotential anomaly and baroclinic currents show the highest variability in the eastern WSC branch, over the continental shelves of Spitsbergen, the Bear Island and the Barents Sea. In the western branch the increased variability occurs at 74°N latitude, above the Mohn Ridge, and in the area where the Atlantic Water recirculates, west of Spitsbergen.

The structure of baroclinic currents, particularly the division between the area of meridional and zonal flow is well shown by the mean values of the U_{30} and V_{30} components at 100 dbar (Fig. 6.2).

Also the July 2000–2007 mean kinetic energy of the baroclinic flow at 100 dbar (Fig. 6.3) clearly shows the diverse intensity of currents in different regions. The most intense flow, forced by the baroclinic effect is situated in the south-western part of the study area, at the bend of the line of underwater ridges. There the mean flow divides into two branches. One of them flows along the line of underwater ridges, whereas the other follows the shelf break. The both currents re-join west of Spitsbergen at the latitude of its central part (ca. 78°N). The convergence intensifies the currents, which is visible in the value of mean kinetic energy. Some intensification is also visible between the Bear Island and the northern Norway in

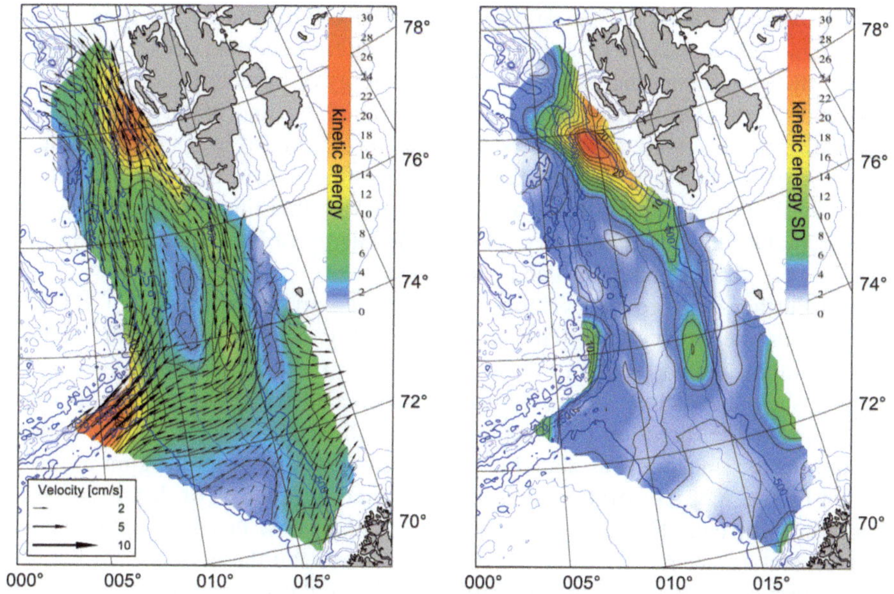

Fig. 6.3 Mean kinetic energy (cm^2/s^2) of the baroclinic flow at 100 dbar shown by the colour scale and calculated mean baroclinic currents (on the *left*) and standard deviation of kinetic energy from July 2000 to 2007

the BSO area. In that case the baroclinic currents are eastward. The NwASC flow along the Barents Sea shelf from the northern Norway to the Bear Island is not visible. There may be different reasons firstly, currents in that area are mostly barotropic, secondly, the horizontal resolution of measurements in that region is low, there are no measurements in the centre of the triangle formed by sections K, H and V1, which may make the image less legible. Thirdly, the figure shows mean currents.

The highest variability of the flow kinetic energy occurs near Spitsbergen, west of the 500 m isobath, with a maximum near the 78°N latitude. All the WSC branches converge here, due to the bottom topography. The highest variability is observed in the eastern branch (core) of the WSC. It is quite an unexpected conclusion and its sources should be probably sought in the extraordinary intensification of flows in the period 2005–2006.

It is worth emphasizing again that the presented currents and the kinetic energy of flows represent only the baroclinic part of those total values. For the AW flow, the barotropic currents are also important and they dominate along the shelf of Norway, the Barents Sea and the western Spitsbergen.

The two branches of flow shown in Fig. 6.3 represent the average from 8 summers. In fact, the flow may have various configurations—it may intensify in the eastern or western branch, or in both branches at the same time. The existence of both branches and the variable intensity of their flow may be significant for the

Atlantic Water inflow to the Arctic Ocean. The bifurcation of the flow is most visible in 2005 and 2006 (Fig. 6.4), a similar pattern occurs in 2000. Attention is drawn by a significant increase in the kinetic energy of baroclinic currents in 2005 and 2006. After the minimum of ca. 4 cm^2/s^2 in 2003 and 2004, the value rose more than twofold—up to 8.5 cm^2/s^2 in 2005 and 9.5 cm^2/s^2 in 2006. Distributions of velocity in 2005 and 2006 indicate a dominating northward flow towards the Fram Strait, thus, a conclusion is reached that during that time the AW inflow into the Arctic Ocean should intensify. The kinetic energy of baroclinic flows dropped 2007 as the WSC temperature and salinity fell.

The further part presents vectors of baroclinic currents at a level of 100 dbar overlaid on the temperature fields at that level (Fig. 6.5). The baroclinic currents were combined with temperature on purpose, not only for plotting reasons. The figures show how close the relationship is between the pattern of temperature isolines and the direction of baroclinic currents. Such a concurrence obviously should occur, since, as was stated earlier, steric relations in that region are dominated by the AW temperature, and the horizontal differences of density generate baroclinic currents. However, the distributions also show that horizontal changes of temperature are mostly of advective nature here, and the coincidence between the pattern of isotherms and the direction of calculated currents show that the importance of the baroclinic currents for mass and energy flux is very high. A very high coincidence may be observed between the varying patterns of the isotherm field and intensification (or weakening) of baroclinic currents.

The series of distributions shows two various types of circulation. The period 2001–2003 features the predominance of transport towards the Barents Sea, 2000 and 2004 are intermediate periods, and in the period 2005–2006 transport towards the Fram Strait prevails. Despite the fact that the transport was still relatively big in 2007, it already began to fall. The tendencies are clearly visible in the time plot in Fig. 6.5. The U_{30} component, which stands for transport towards the Barents Sea and is marked as the red line, reaches its maximum in 2002. The V_{30} component begins to dominate after 2004, and the years 2000 and 2004 are the period of equilibrium between the both advection tendencies.

In the period from 2005 to 2006 an extraordinary northward expansion of warm AW took place. The 5 °C isotherm at a level of 100 dbar moved at a velocity of more than 2° of latitude per year. The isotherm had crossed the 76°N latitude before, however, those were usually detached, discontinuous structures. In 2005 a wide tongue of the AW with temperature 5 °C crossed that latitude, and in 2006 it reached 78°30′N. In 2007 the meridional extension of the isotherms at 100 m returned to the distribution from 2005. Those phenomena will be discussed further herein.

The time series of currents in Fig. 6.5 shows the similarity of the mean current velocity and the component V_{30}. Actually, high correlation coefficients between the velocity or kinetic energy of currents and the V_{30} component indicates domination of flows towards the Fram Strait (Table 6.1). The correlation between the components of the flow is low. However, the significant negative correlation between the components U_{30} and V_{30} with 1 year lag: U_{30} (2000–2006)—V_{30} (2001–2007), $r = -0.71$ is worth consideration. It means that the following year after the

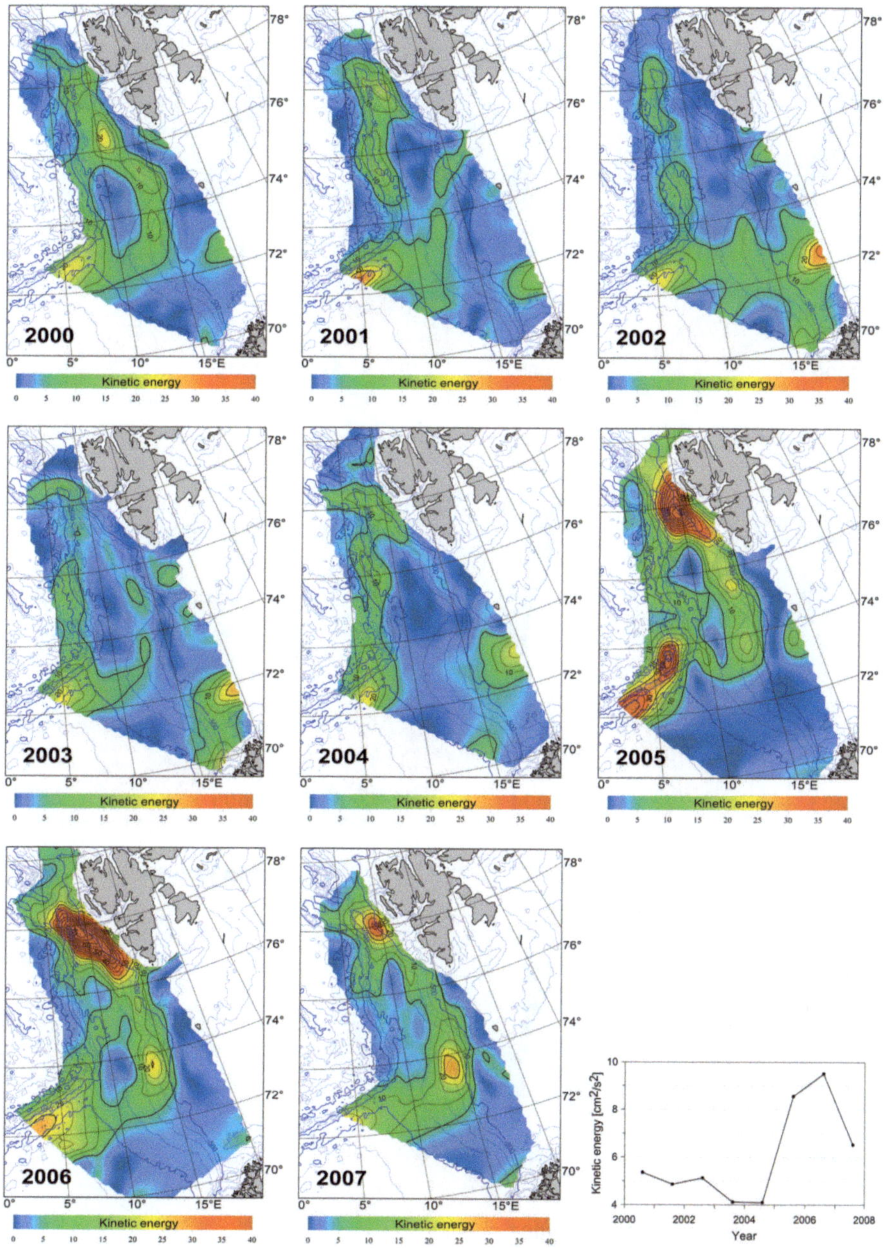

Fig. 6.4 Distributions of the kinetic energy (cm^2/s^2) of flow at 100 dbar in subsequent summer seasons and time series of the mean kinetic energy at 100 dbar from the AREX study area. The 5 and 40 cm^2/s^2 isolines are bold

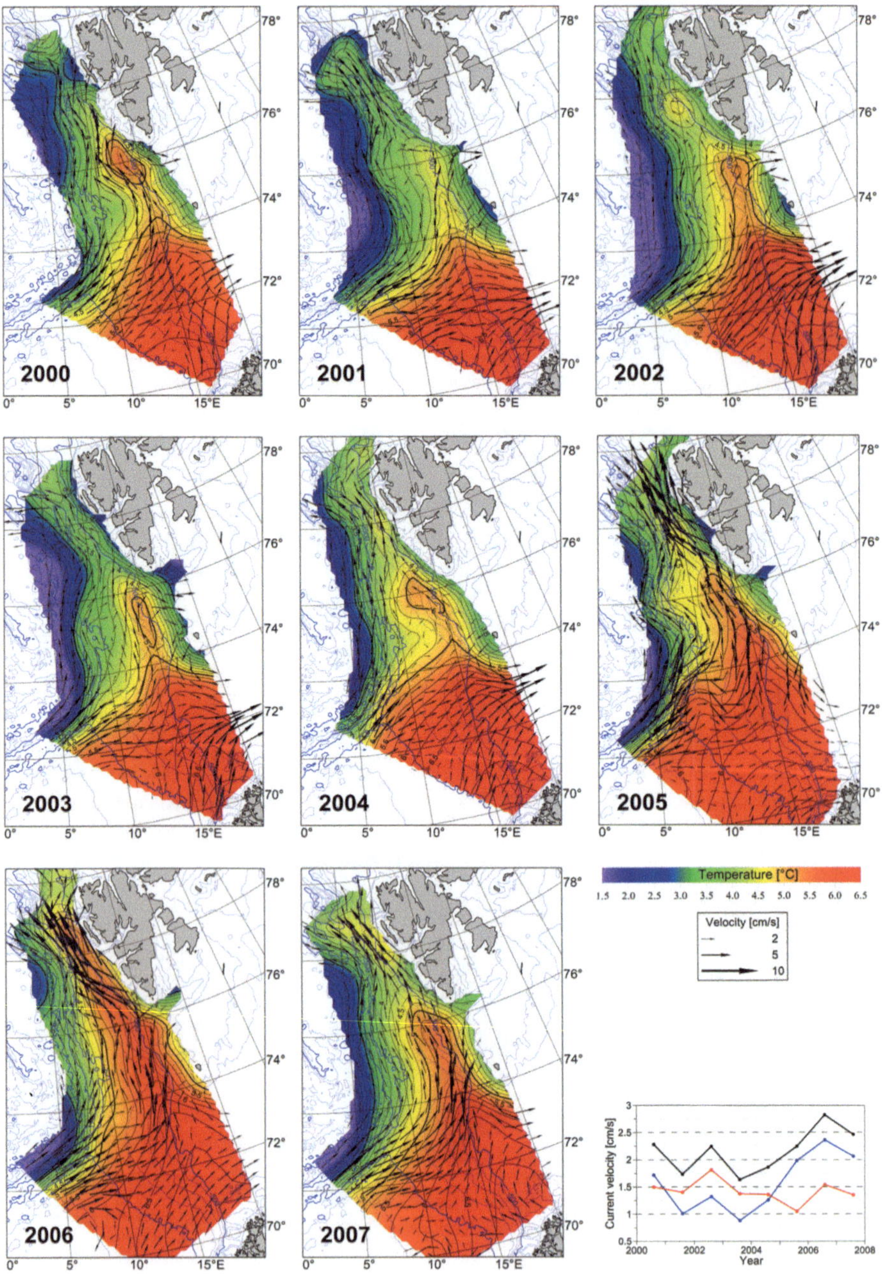

Fig. 6.5 Water temperature (*colour scale*) and baroclinic currents (*vectors*) at 100 dbar in July 2000–2006. The 3 and 5 °C isotherm is bold. The line plot shows time series of mean current velocity (*black line*), the U_{30} component (*red line*) and the V_{30} component (*blue line*)

Table 6.1 The correlation coefficients between the basic dynamic properties at 100 dbar for the entire AREX study area. The data are detrended

Velocity	U	V	U_{30}	V_{30}	KE
Vel	0.42	**0.99**	0.48	**0.92**	**0.74**
U		−0.53	0.59	**0.73**	−0.69
V			0.37	**0.97**	**0.76**
U_{30}				0.11	−0.02
V_{30}					**0.82**

Table 6.2 The correlation coefficients r between the temperature at 100 dbar and the AW layer temperature, and current velocities at 100 dbar for the entire AREX study area

Temperature	Vel	U	V	U_{30}	V_{30}
T_100	0.54	−0.45	0.56	0.04	0.59
T_AW	0.63	−0.65	0.69	−0.06	**0.75**

intensification of flux into the Barents Sea one can expect a decrease of the baroclinic flow towards the Fram Strait and mean current velocities in the WSC.

Next, time series representing the dynamics of baroclinic currents are compared to time series of temperature of the entire AREX study area. This comparison was performed for a temperature at 100 dbar and mean temperature of the AW layer (Table 6.2). As in other cases, the data were detrended.

The results show that correlations with averaged temperature of the AW layer are more significant than those with the temperature at 100 dbar. That is easy to understand, since the geopotential anomaly, thus baroclinic currents, is closely related to the temperature in the entire water column above the NML. The highest correlation, $r = \mathbf{0.75}$, was obtained for the component V_{30} which represents the flow along the WSC axis. The lack of correlation for the component U_{30} is surprising. Instead, a significant correlation with 1 year lag (not shown in the table) occurs, between U_{30} (2000–2006) and T_{AW} (2001–2007), $r = -\mathbf{0.88}$. It means that the AW temperature decreases with 1 year time lag after intensification of flow into the Barents Sea. It is probably the same mechanism that occurs for the above-described negative correlation between U_{30} (2000–2006)—V_{30} (2001–2007), $r = -0.71$. An attempt may be made to explain that with a simple mechanism—an increase of transport into the Barents Sea (higher U_{30}) means a decrease in northward warm water transport, towards the AREX study area (lower T_{AW} and weaker northward currents the following year).

For the value of U_{30} (2000–2006) and T_{100} (2001–2007) there is a similar significant correlation, $r = -\mathbf{0.88}$. It is clear that there is feedback between the AW temperature and baroclinic flow, and not only temperature modifies the flow by intensification or weakening of the baroclinic effect, but the flow influences temperature, thus, the baroclinic effect, too. This is confirmed by correlations between the AW temperature and the sea region dynamics, calculated separately for the southern and the northern part.

6.3 Division of the Study Area

The results of the study concerning the baroclinic currents field confirm the previous conclusion that it is justified to divide the region into two parts: an area where flows into the Barents Sea dominate and the area of intense advection towards the Fram Strait. The plots of the currents velocities in those regions show different time series (Fig. 6.6). In the case of the southern part, the average velocity of baroclinic currents is determined by the U_{30} component, thus, the flow into the Barents Sea. The velocity maximum was observed in 2002, the minimum—in 2005. The V_{30} component varies similarly but velocities are lower. For the northern part, the V_{30} component coincides mostly with the mean velocity. The flows across the WSC (component U_{30}) are steady, at a level of 0.5 cm/s. For currents towards the Fram Strait, the maximum is found in the period 2005–2006 and the minimum—in 2003. A wide range of change in mean baroclinic velocities in that region draws attention.

Correlations were calculated for a few cases, while considering relationships between different values: for the entire region, for the southern part, for the northern part and mixed correlations in various configurations. The autocorrelation with 1 year lag was also calculated. Large values of the correlation coefficient for the U_{30} component in the southern part confirm the domination of zonal flows (Table 6.3). Correlation between the U_{30} and V_{30} current components is higher than for the entire study area. For the northern study area, the correlation coefficient of the V_{30} component with the current velocity clearly indicate that meridional flows prevail (Table 6.4).

There are no significant correlations in the southern part, however, the northern part reveals important correlations between the fluxes along the WSC and the AW layer temperature (Table 6.5). That may mean that the northward heat flux in this part is dominated by the baroclinic currents. However, 1-year time lag of temperature in relations to the sea region's dynamics for the southern part is visible. The results are so interesting that they are presented in a separate table (Table 6.6).

The AW layer temperature decrease in the southern part correlates with the flow increase in the preceding year. A decision was made to calculate once more the

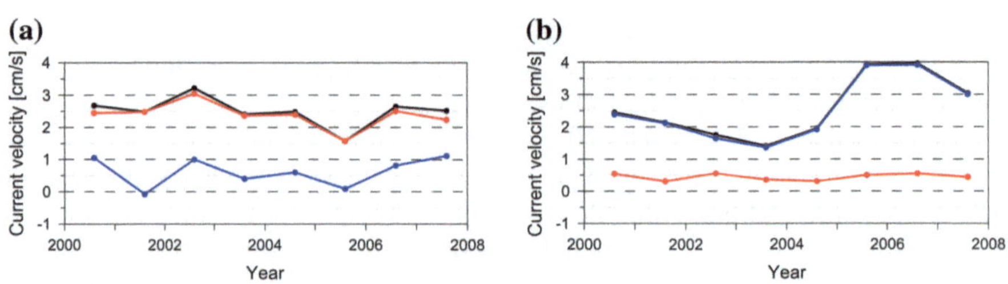

Fig. 6.6 Time series of mean currents velocity (*black line*), the U_{30} component (*red line*) and the V_{30} component (*blue line*) for the southern study area, below the latitude 74°N (**a**) and the northern study area (**b**)

Table 6.3 The correlation coefficients r between velocities at 100 dbar in the southern part of the AREX study area. The data are detrended

Velocity	U_{30}	V_{30}
Vel	**0.98**	0.73
U30		0.58

Table 6.4 The correlation coefficients r between velocities at 100 dbar in the northern part of the AREX study area. The data are detrended

Velocity	U_{30}	V_{30}
Vel	0.57	**0.999**
U_{30}		0.55

Table 6.5 The correlation coefficients r between the AW layer temperature and the currents velocities at 100 dbar in the southern and northern part of the AREX study area

Temperature	Vel	U_{30}	V_{30}
T_{AW} south	0.02	−0.01	0.18
T_{AW} north	**0.84**	0.53	***0.84***

Table 6.6 The correlation coefficient r between the mean AW layer temperature from July 2001–2007 and the dynamics of baroclinic currents in the period 2000–2006 at the level of 100 dbar in the southern part of the AREX study area

Temperature	Ke_south (2000–2006)	Vel_south (2000–2006)	U_{30}_south (2000–2006)	V_{30}_south (2000–2006)
T_{AW} south (2001–2007)	**−0.93**	**−0.86**	**−0.80**	**−0.78**

mixed correlations, between T_{AW} in the southern part and the currents velocities in the northern one, and vice versa, between the T_{AW} in the northern part and the currents velocities in the southern part. Although the first operation yielded no meaningful results, the second one confirmed the previous assumptions. The values of concurrent correlations are not significant, however, correlations with 1 year time lag give significant results (Table 6.7).

Negative correlations were found again between the current dynamics in the southern part and the AW layer temperature values (this time in the northern part)

Table 6.7 The correlation coefficient r between the AW layer mean temperature in 2001–2007 in the northern part of the AREX study area and the dynamics of the baroclinic currents in years 2000–2006, at the level of 100 dbar in the southern part of the AREX study area

Temperature	KE_south (2000–2006)	Vel_south (2000–2006)	U_{30}_south (2000–2006)	V_{30}_south (2000–2006)
T_{AW} north (2001–2007)	**−0.81**	**−0.82**	**−0.80**	−0.54

in the following year. The current dynamics in the southern part represents mostly the inflow into the Barents Sea. Thus, a conclusion may be finally formulated that the increased baroclinic AW flux into the Barents Sea correlates with the AW temperature decrease in the entire AREX study area in the following year. It is impossible to state directly that it is the flow into the Barents Sea that causes the decrease in temperature; both of the factors may be caused by the same forcing. However, in the case of temperature decrease in the northern part, it makes a logical sequence: the eastward transport increase reduces the northward warm water flow and the decrease of the WSC temperature.

6.4 In Situ Measured Currents

The Institute of Oceanology gathered a vast data base of measurements collected with the ADCP mounted in the hull of the R/V Oceania. Measurements with use of LADCP fixed to the bathymetric rosette have been a standard operation since 2003. These results enrich our knowledge and show complex dynamics of the studied environment. The first impression from the analysis of current vectors, obtained with the above methods and drawn on the map gives an impression of a chaotic system. It is mainly due to the fact that, unlike the hydrographic properties (temperature, salinity), time scale of changes in the sea region's dynamics is much shorter than the duration of a cruise. Thus, the results from current profilers must not be taken as synoptic data—the values cannot be interpolated onto a map to obtain a synoptic image. It happened in the past that measurements taken at the same station with intervals of several hours revealed currents flowing in different directions (Walczowski et al. 2005). Furthermore, polar regions are characterized by a small Rossby radius of deformation; in this particular area it is of approx. 11 km. It means that in a distance of 11 km currents may flow in opposite directions (Figs. 6.7, 6.8).

High temporal of flows is caused (apart from the tides which are filtered) by the barotropic component (constant with depth) of currents. The baroclinic and barotropic component are always superimposed (Fig. 6.9), "clean" baroclinic or barotropic currents are a rare case.

The data were gathered during Arex 2005 and Arex 2006 LADCP measurements. The red colour denotes the U component of the current (eastward), the green colour—the V component of the current (northward).

The region of the Barents Sea shelf and the slope of the western Spitsbergen creates particularly favourable conditions for occurrence of sea surface tilt by the influence of wind and formation of barotropic currents. Varying winds may modify the direction of sea surface tilt and, thus, the direction of flow. A barotropic signal propagates with the speed of sound (ca. 1500 m/s in water). In the WSC core, the direction of flow is always along the shelf, only the velocity vector sense may change from "towards" to "from" the Fram Strait. The northern currents towards

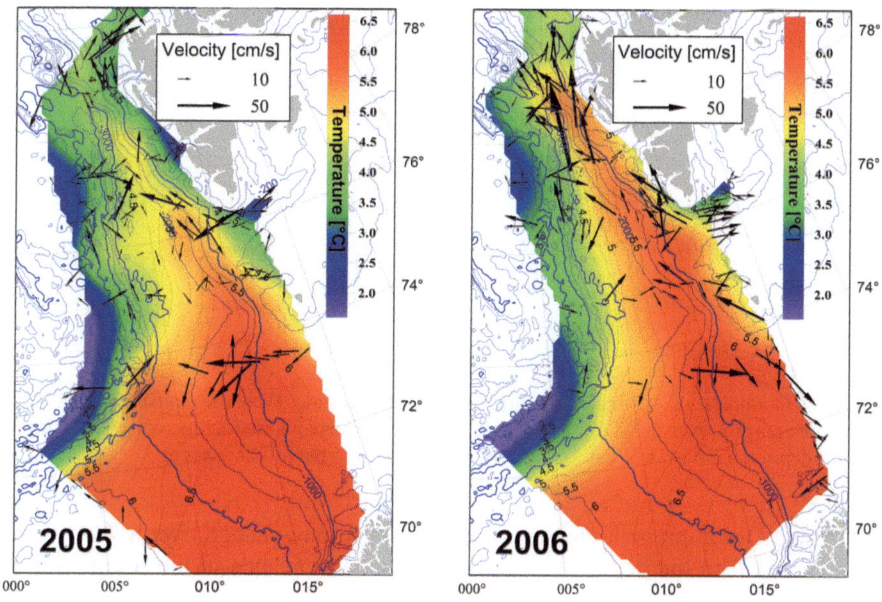

Fig. 6.7 Temperature (*colour scale*) and vectors of currents measured with LADCP at a depth of 200 m

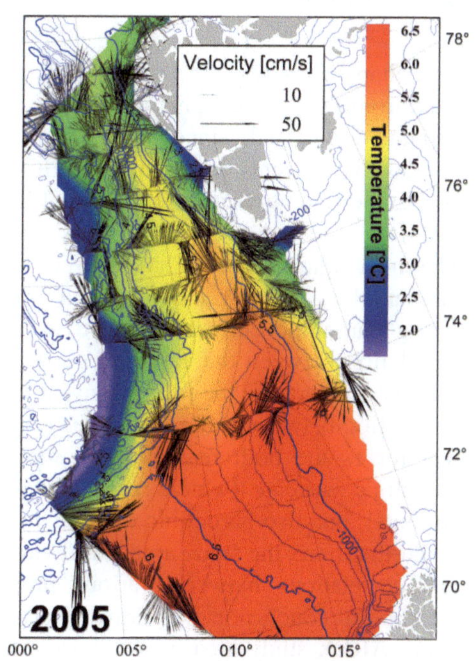

Fig. 6.8 Temperature (*colour scale*) and vectors of currents measured with VmADCP at a depth of 100 m in June 2005

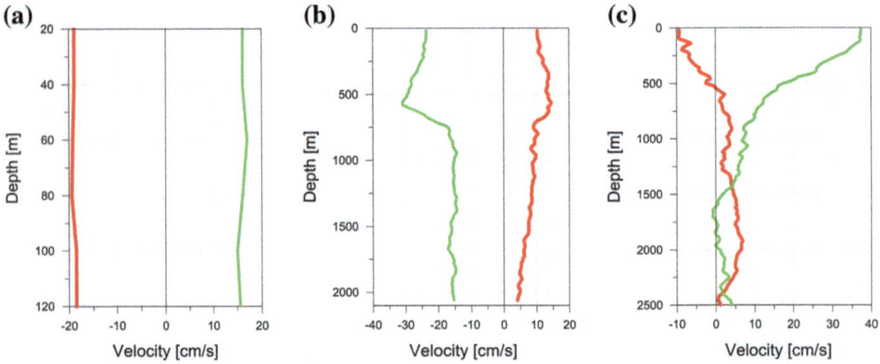

Fig. 6.9 Examples of vertical structure of current near the Western Spitsbergen. **a** Clean barotropic current over the shelf. **b** Strong barotropic currents and baroclinic current overlapped from a depth of 750 m. **c** Weak barotropic flux and strong baroclinic current

the Arctic Ocean prevail. However, sometimes it may change over a short period of time and then the negative barotropic component results in the southern WSC transport.

6.5 Summary

This chapter describes the sea currents in the studied area as measured or calculated from the hydrographic data. To analyse the dynamics, mostly the baroclinic currents were used, since the hydrographic data form a coherent time series, besides, the currents represent the THC best. Periods of intense baroclinic transport to the Barents Sea and a period of intense flow towards the Fram Strait (after 2004) were recognized. Correlations between various properties were calculated. The most significant correlations were obtained between the AW mean temperature and the currents velocities. High temperature of the AW layer correlates with strong baroclinic currents directed towards the Fram Strait. A strong negative correlation between the eastward current velocity (towards the Barents Sea) and the AW temperature the following year is even more interesting. If there is a direct connection between those properties (and not a correlation with another property that forces the both reactions), it shows that not only temperature determine baroclinic currents (which results from the principle of calculating baroclinic currents), but also baroclinic currents impact the temperature field.

Furthermore, the study area was divided into the northern and the southern part, and correlations between various properties were analysed. Finally, a very meaningful conclusion was drawn that a decrease of the AW temperature in the entire AREX study area is correlated with an increase of the baroclinic currents velocity towards the Barents Sea in the preceding year.

References

Fahrbach E, Meincke J, Østerhus S, Rohardt G, Schauer U, Tverberg V, Verduin J (2001) Direct measurements of volume transports through Fram Strait. Polar Res 20:217–224

Hanzlick DJ (1983) The West Spitsbergen Current: transport, forcing and variability. Ph.D. thesis, University of Washington, Seattle

Osinski R, Wieczorek P, Beszczynska-Moller A, Goszczko I (2003) ADCP-referenced geostrophic velocity and transport in the West Spitsbergen Current. Oceanologia 45(3):425–435

Rudels B (1987) On the mass balance of the Polar Ocean, with special emphasis on the Fram Strait, Norw. Polar Inst., 188, 53.

Schlichtholz P, Houssais MN (1999) An inverse modeling study in Fram Strait part I, part II: dynamics and circulation. Deep-Sea Res II 46:1083–1168

Walczowski W, Piechura J, Osinski R, Wieczorek P (2005) The West Spitsbergen Current volume and heat transport from synoptic observations in summer. Deep-Sea Res I 52:1374–1931

Chapter 7
The West Spitsbergen Current Volume and Heat Transport

7.1 Meridional Variability

Transport of heat by a sea current depends on the volume transport and water temperature:

$$Q = \int_S c_p \rho v (T - T_{ref}) ds \qquad (7.1)$$

where ρ is water density, c_p is water specific heat (3,985 J kg^{-1} K^{-1}), v is velocity perpendicular to the area s, T is water temperature, T_{ref} is reference temperature. In the case of the WSC and the Arctic Ocean balance, the reference temperature of −0.1 °C is conventionally adopted, however, more and more opinions occur stating that a different value should be applied (Schauer et al. 2008). It seems that using sea water freezing temperature would be the most reasonable. However, in the present paper the value of −0.1 °C has been assumed for T_{ref} in order to be able to compare the results with other publications.

The values of total transport and the Atlantic Water transport have been calculated for all the sections. Similarly to Chap. 5 where the variability of mean AW temperature and salinity in function of latitude is presented, the current section presents the change in the baroclinic transport of the AW with latitude first (Fig. 7.1). The volume transport changes from 2.5 Sv at 75°N to 0.6 Sv in the Fram Strait, and the heat transport from 42 to 12 TW, respectively.

A disproportionately low transport through the southernmost section situated along the parallel 73°30′N (transect H) can be easily observed. It is located in the southern part of the study area, whereas the other transects are in the northern one (according to the previously introduced division). The largest transport occurs through the parallel 75°N and then it falls down as latitude increases. The same is observed for heat transport. In that case the exceptionally low heat transport across transect H results from the low volume transport, since the AW temperature at that transect is predictable (Chap. 5). The observed decrease of transport at the latitude of 73°30′ may have various reasons. The fact that the section is aligned askew to the flow should not be of any importance (we calculate only the perpendicular

W. Walczowski, *Atlantic Water in the Nordic Seas*,
GeoPlanet: Earth and Planetary Sciences, DOI: 10.1007/978-3-319-01279-7_7,

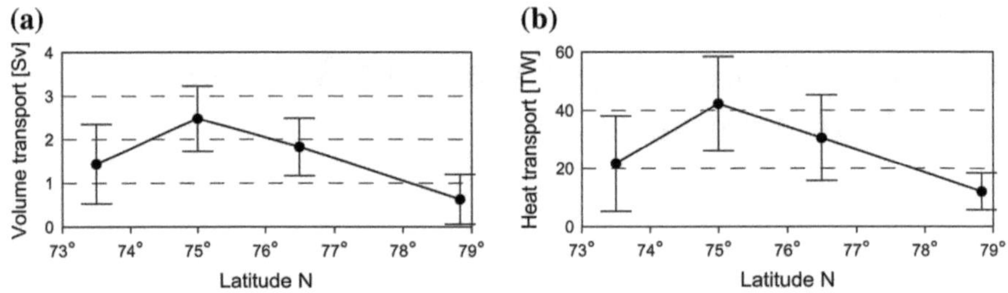

Fig. 7.1 **a** The July 2000–2007 mean baroclinic volume transport of the Atlantic Water (Sv) and **b** heat transport (TW) through subsequent parallel sections. Standard deviations are marked

component, but when the current flows at an angle to the transect, the field of the considered flow increases proportionally). The most rational explanation of that lower transport seems to be the fact that transect H does not close the entire AW domain. From the west, the transect is crossed by the Arctic Front, however, in the east it ends at a depth of 400 m, in the Bear Island Trough, which is the deepest BSO region. The author performed calculations for analogical transects using data from numerical model (Maslowski et al. 2000). A similar phenomenon of significant decrease of transport across the transect 73°30′N was observed. When the present paper was under preparation, the AW recirculation in the Barents Sea was described (Skagseth 2008). A part of the AW flowing through the southern part of the BSO recirculates along the 250 m isobath and leaves the Barents Sea through the northern part of the BSO. It was estimated that approximately 0.9 Sv of the AW flows westward through the Bear Island Trough, bypassing transect H. To some extent that explains the above-mentioned deficiency of volume and heat transport. Another independent cause of the decreased transport may be the fact that in the southern part of the described sea region flows are more barotropic than in the northern one, thus, the calculations cannot taken them into account. Zonal gradients of temperature and hence the gradients of geopotential anomaly rise with latitude, thus the baroclinic currents are more intense in the north. It may be caused by the WSC core, as it was mentioned before that the temperature in the core hardly changes in the northern section. With temperature decreasing westward from the shelf break, that must intensify the baroclinic flow. No such feature is observed in the southern part, thus, the baroclinic currents are not so intense. It is visible in the properties of the WSC core that the farther north the larger the maximum flow velocities are found. Anyway, it is distinct that dividing the study area into the northern and the southern part was very much justified since they feature totally different dynamics.

By using the linear extrapolation of data from Fig. 7.1, one can estimate that corrected values for the AW northward volume and heat transport through 73°30’ are 3.6 Sv and 60 TW, respectively. Recirculation of the AW and “bypassing” the flow by the AW amounts to ca. 0.9 Sv. The remaining 1.3 Sv comes probably from the barotropic flows.

7.2 Temporal Variability

Chapter 5 presented the mean and temporal variability of hydrographic properties on subsequent transects across the WSC. The author will focus in this part on dynamic properties (volume transport, heat transport, flux structure), although it will be impossible not to refer to physical properties as they are inseparably connected with baroclinic flows. All the positive flows and transport on zonal transects are northward, and on the meridional transect V1—eastward.

7.2.1 *Transect H*

The transect shows intensification of the baroclinic flow above the underwater ridge and above the Barents Sea slope (Fig. 7.2). Interpreting the magnitude of flows across that transect is problematic, since it is situated in the area where the flow divides into the eastern and northern branch. Mean flows are lower than those on transect K (Fig. 7.3) which is situated north of the above-mentioned transect. Similarly, lower values of transport across transect H and K are given by Piechura et al. (2001). Time series shows a significant decrease of the AW transport from 2000 to 2004 and then an increase until 2007.

7.2.2 *Transect K*

The transect is situated along the parallel 75°N and it features distinct northward baroclinic flows (Fig. 7.3). Two areas of the flow can be distinguished: the western WSC branch on the eastern side of the Mohn Ridge and the WSC core above the Barents Sea slope. The baroclinic flow velocities vary between −17 cm/s (southward flow) and 26 cm/s. Two maxima are distinct in the time series: in 2000 and 2006. A sudden decrease of the AW volume and heat transport occurred in 2007.

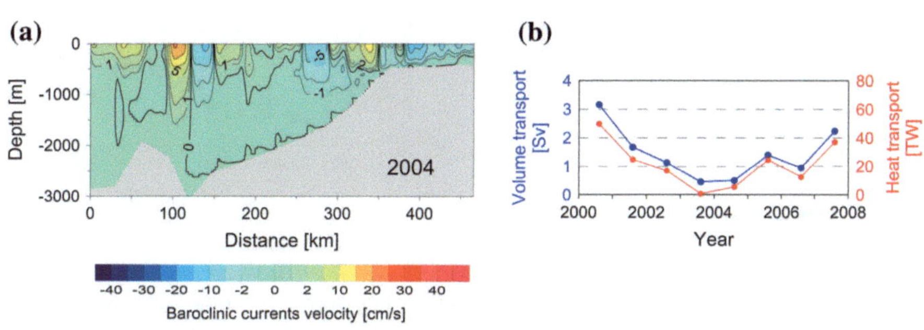

Fig. 7.2 The baroclinic flow in July 2004 (**a**) and the July 2000–2007 time series of baroclinic volume transport and heat transport of the Atlantic Water layer (**b**) across transect H

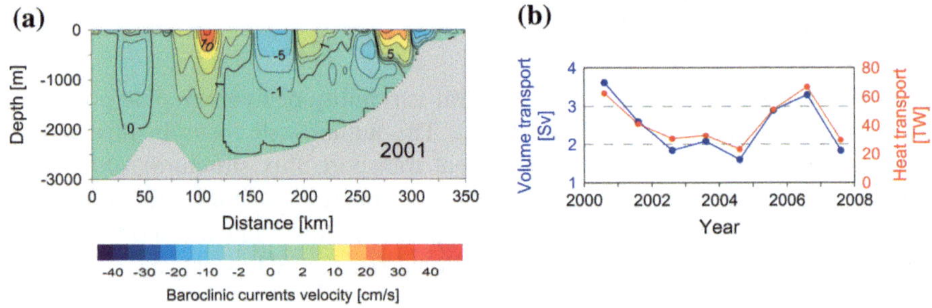

Fig. 7.3 The baroclinic flow in July 2001 (**a**) and the July 2000–2007 time series of baroclinic volume transport and heat transport of the Atlantic Water layer (**b**) across transect K

7.2.3 Transect N

The structure of flow across the parallel of 76°30′, where transect N is situated, is similar to a structure at the previous transect: there is a deep flow above the underwater ridge and a more shallow and more intense flow above the Spitsbergen shelf slope (Fig. 7.4). Due to the changing bottom topography, where the ridge approaches the Spitsbergen shelf, the distance between the both WSC branches is reduced to 120 km. The variability of flow is also similar to that from transect K.

The AW volume and heat transport across transect N is highly correlated with the AW temperature in the entire study area. The correlation is even higher than for the AW temperature. The correlation between transports across the transect N and the mean values of component V of the baroclinic current at 100 dbar for the entire study area is also high, below the level of significance of 0.05. It means that during the period of study the transect was to some extent representative for mean values of the AW properties and dynamics in the AREX study area. The author would not like to go into deeper speculations here as to what is the cause or what is

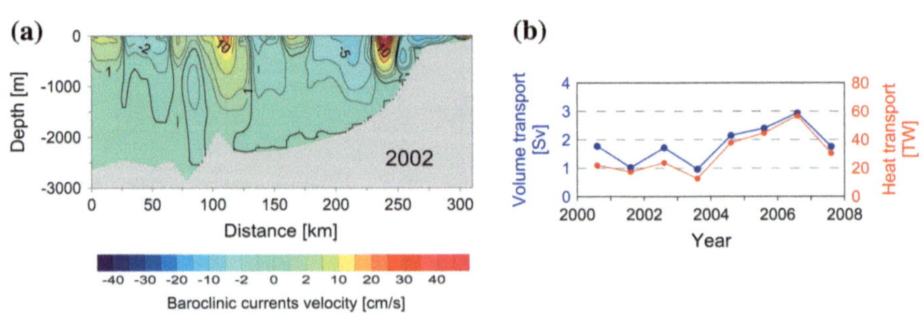

Fig. 7.4 The baroclinic flow in July 2002 (**a**) and the July 2000–2007 time series of baroclinic volume transport and heat transport of the Atlantic Water layer (**b**) across transect N

Table 7.1 Coefficients of correlation r between the AW volumc and heat transport across transect N in July 2000–2007 and the selected AW properties and mean baroclinic fluxes at 100 dbar

Transports	T_{AW} transect N	T_{AW} study area	U study area	V study area	U_{30} study area	V_{30} study area
Volume transport AW (N)	**0.88**	**0.96**	−0.49	**0.73**	0.15	**0.73**
Heat transport AW (N)	**0.89**	**0.97**	−0.49	0.62	0.05	0.64

the effect of described correlations—whether the current dynamics is more intensive due to higher temperature, or the other way round. Locating the transect along the parallel 76°30′N seems appropriate; it is downstream of the area of exchange with the Barents Sea, yet still upstream of the area of the strongest AW recirculation and the Arctic Front bifurcation. A significant question is whether the dependence also occurred in the past and whether it will continue. In the first case it would be possible to reconstruct the variability in the studied region several years back (transect N was explored longer than the entire study area), and in the second case, conclusions could be drawn about the mean conditions of the entire region, without the necessity of the full set of in situ measurements, on the basis of studies performed on transect N alone. It is worth noting that the author is not a strict advocate of the idea of "representative points", based on which we can diagnose on the state of the entire sea region, however, the author claims that some solutions should be sought to make is possible to draw more general (yet cautious) conclusions (Table 7.1).

Figure 7.4 clearly shows that the heat transport across the transect is strongly related to volume transport (r = **0.93**) and that the increase of the baroclinic heat transport towards the Fram Strait occurred mostly due to intensification of currents (which, on the other hand, is related to a positive correlation with temperature, thus, a temperature rise also matters). For time series extended since 1996 (Fig. 7.5), the coefficient of correlation between the heat transport and the flow (after detrending) continues to rise and equals r = **0.97**. The coefficient of correlation of the heat transport–temperature pair is much lower: r = **0.63**. On the other hand, the series of volume and heat transport for that period show that the volume transport does not have a clear trend, mostly due to the large transport in 1996.

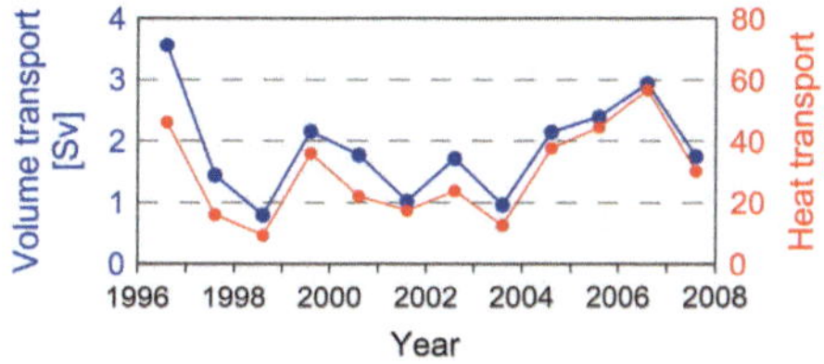

Fig. 7.5 Transect N. Volume transport and heat transport in the period 1996–2007

7.2.4 Transect EB

That is the northernmost transect at which IOPAS performs systematic measurements. For the given period of study, the mean AW volume and heat northward transport is 0.64 Sv and 12.6 TW, respectively. Two baroclinic AW flows are also visible in that transect: the Svalbard Branch flowing above the Spitsbergen shelf and the Yermak Branch above the Knipovich Ridge (Fig. 7.6). The maximum of the baroclinic transport into the Arctic Ocean was observed in the period 2000–2001 and 2004–2005, the minimum in 2003.

The correlation between the AW temperature on transect EB and N is r = **0.74**, for salinity r = 0.63 (Chap. 5). The correlation between the volume transport across transects is lower, r = 0.52, for heat transport it rises (probably due to better correlation of temperatures) to r = **0.73**. Thus, an important questions occurs: to what extent is transect EB representative for transports through the Fram Strait into the Arctic Ocean? The analysis of horizontal distributions shows that the flows above the parallel of 78°N are highly variable, they are rinfluenced by the processes occuring at the edges of deep basis, located to the north and south of this area. Similar results were obtained by the analysis of data gathered by isopycnal floats, performed by the author (not included herein). Thus, one can state that in the case of measurements from a ship, transect N may be more representative for mean flows through the Fram Strait, since it is less susceptible to current variations on short time scales.

7.2.5 Transect V1

The transect along the meridian of 20°E represents the AW transport into the Barents Sea. The structure and temporal variability of that current is completely different from the previously presented flows towards the Fram Strait. The AW inflow occurs along the entire width of the BSO, with intensification in the

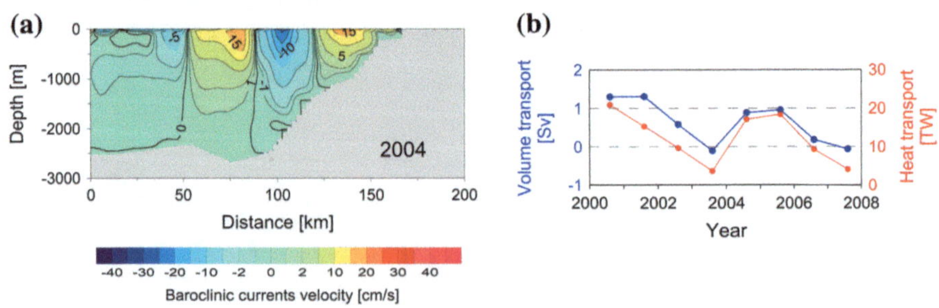

Fig. 7.6 The baroclinic flow in July 2004 (**a**) and the July 2000–2007 time series of the AW baroclinic volume and heat transports (**b**) across transect EB

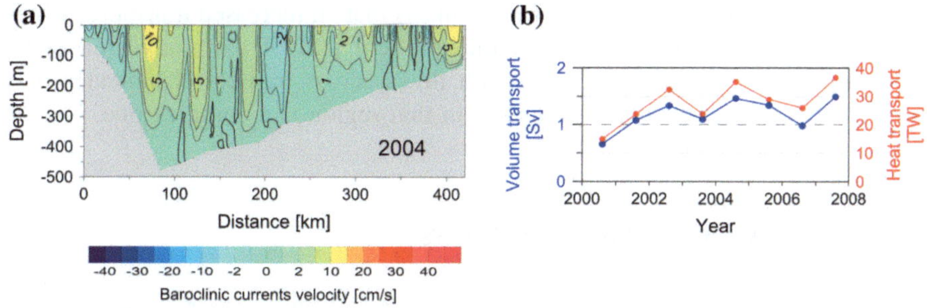

Fig. 7.7 The baroclinic flow in July 2004 (**a**) and the July 2000–2007 time series of baroclinic volume transport and heat transport of the Atlantic Water layer (**b**) across transect V1

Table 7.2 The coefficient of correlation r for mean AW properties on transect V1 with transects H and K

Transect	T_AW	S_AW	Vol_transport	Q_transport
H	0.31	**0.99**	−0.36	−0.26
K	0.10	**0.80**	−**0.87**	−**0.85**

northern part (Fig. 7.7). The mean transports of the AW volume and heat in the period of study are 1.28 Sv and 28.8 TW, respectively. The values are more than twice as high as the values of transport into the Arctic Ocean across transect EB2. Most of the flows through the BSO in form of barotropic currents, strongly influenced by the wind field. The mean transport from the period 2003–2005, calculated on the basis of measurements with current meters, equals 1.1 Sv, composed of the inflow of 2 Sv and the outflow of 0.9 Sv (Skagseth 2008). The value is very close to the calculated annual baroclinic flows presented herein. Unfortunately, the variability in time does not correspond entirely; Skagseth specifies the maximum of volume transport in summer 2005 at the level of 4 Sv.

The relationships between the physical properties and flows of the AW on the nearest zonal transects and the meridional transect V1 are meaningful (Table 7.2). There is a very high correlation between the salinity on transect H and V1 and a bit lower, although still at the level of significance 0.05, between transects K and V1. It seems obvious, since close proximity of the transects justifies such correlations. However, they do not occur for temperature; the AW temperature at transects H and K is not correlated with temperature at transect V1. It means that AW temperature at the transect V1 must undergo significant modification. That confirms the thesis on recirculation of the AW in the Barents Sea. A portion of the AW in the Bear Island Trough is colder due to circulation along a path of approximately 700 km.

The correlation between the transports across transect K and transect V1 is much more significant. For the volume transport the coefficient of correlation equals −**0.87**, and for the heat transport: −**0.85**. It can be clearly seen that an increase of the AW transport into the Barents Sea is accompanied by the decrease of northward transport. That confirms the thesis that an increased transport into the Barents Sea reduces temperature in the northern part of the study area in the following year. Lack of correlation with flows across transect H indicates once more the complex structure of transport in that region is a complex one.

7.3 Baroclinic vs Barotropic Flows

Comparison of currents measured *in situ* and those calculated from hyrographic data enables calculating the barotropic component from the ADCP data. It can be also calculated from the entire current profile measured with the LADCP. The barotropic component may be used as the reference velocity in the mixed method of calculating geostrophic currents (Piechura et al. 2002; Osiński et al. 2003). Walczowski et al. (2005) showed that the structure of baroclinic currents and currents calculated from the LADCP data is similar on transects (Fig. 7.8) and the transports calculated from the LADCP measurements are usually much larger than the baroclinic ones. It is not surprising since the current of 1 cm/s at a transect of 100 km^2 area results in transport of 1 Sv, and mean velocities of barotropic currents are usually around a few cm/s.

In the case of transect N in 2005, a dominating influence of the barotropic component is visible above the continental slope. In summer 2006 there is a very strong barotropic flow at the 200th km of the transect, near stations N0 and N1 (011°E and 012°E) and above the shelf.

The flow structures are well pictured by the plots showing vertically integrated current velocities:

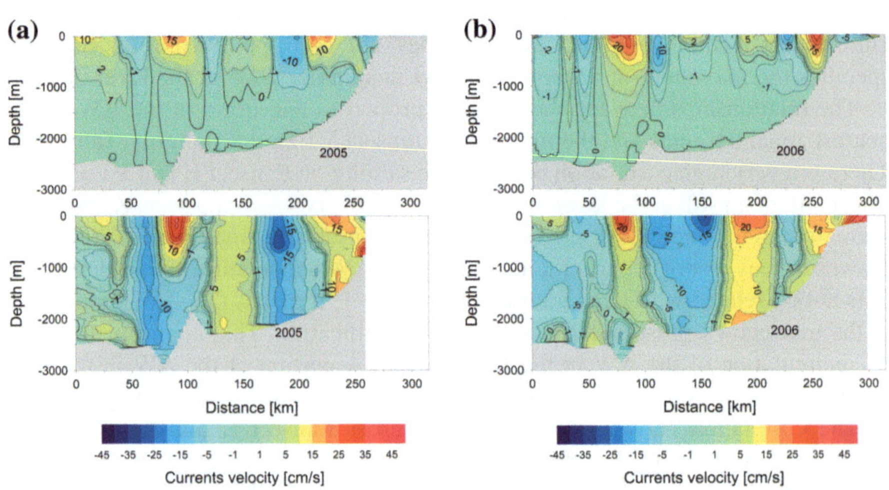

Fig. 7.8 Transect N along the parallel of 76°30′N, summer 2005 (**a**) and 2006 (**b**). The calculated baroclinic currents (*upper images*) and currents measured with the LADCP

$$V\,\mathrm{int}(x) = \sum\nolimits_{-1,000}^{0} V\Delta z. \tag{7.2}$$

The obtained value—transport per a unit of length of transect—is expressed in m^2/s and gives the volume transport after multiplying by the length of the transect. In the case of the presented plots (Fig. 7.9) only the AW transport was calculated, by defining it with temperature (T > 0 °C) and salinity (S > 34.92). Baroclinic velocities calculated from the hydrographic data (black lines) were used as well as

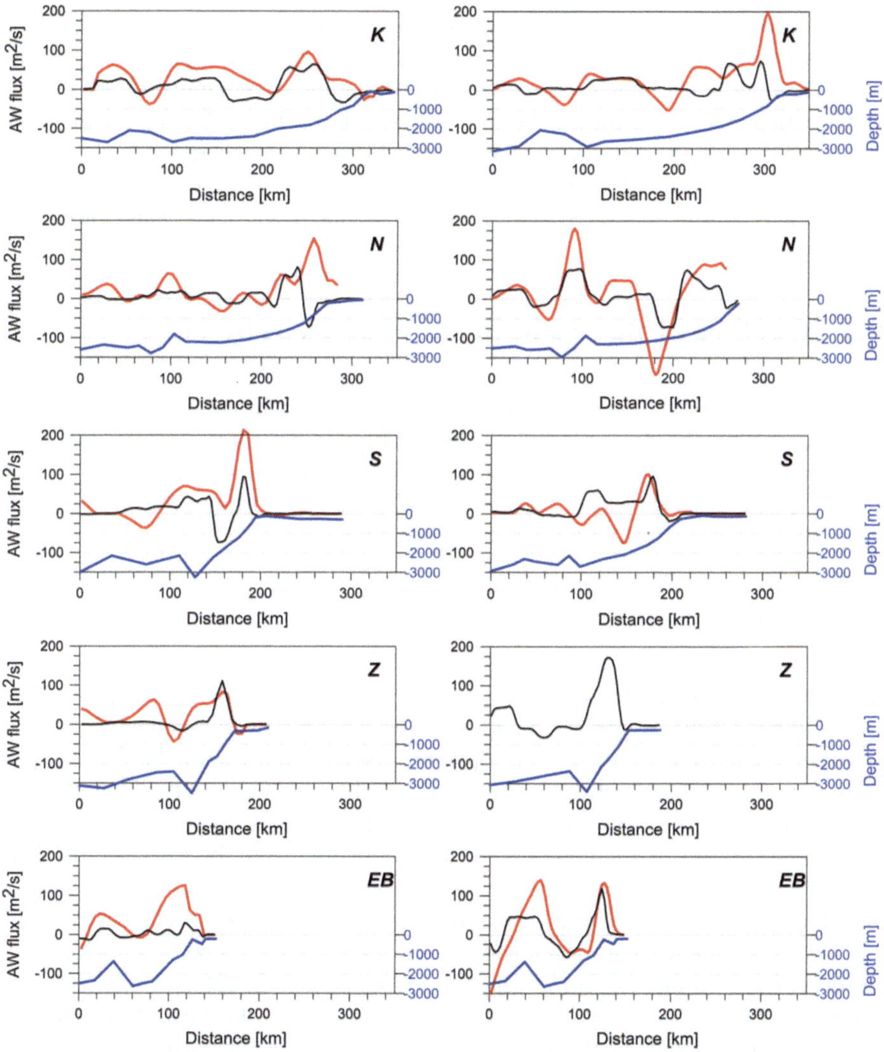

Fig. 7.9 Vertically integrated AW currents on transects of the AREX study area in 2003 (*the left column*) and 2005 (*the right column*). The *red line* denotes the LADCP data, the *black one*—baroclinic calculations. The *bottom line* is denoted by the *blue line*

velocities calculated from the LADCP data (red lines). Only the component perpendicular to the transect was used. The bottom profile (blue lines) on transects help to locate the maxima of flows. Evolution of the flow may be seen in both years from wide and less intense (transect K, 75°N) to the narrowing of the AW flow and convergence of both branches (transects S and Z). In 2003 a difference between baroclinic flow and total flows measured with the LADCP can be clearly seen above the continental slope where the intensity of barotropic flows is significant. The differences are much smaller above deep areas of the region. The centre of the barotropic flow is located above the 1,000 m isobath. It is interesting that in 2005 the transport above the shelf (the eastern WSC branch) was mostly baroclinic, judging by the small difference between the calculated and measured transports The mesoscale activity was more intense in 2005, the anticyclonic eddy on transect N is particularly interesting. It is also well visible on Fig. 7.8. Two distinct AW flows can be seen in both years at the transect EB which is located at the latitude of 78°50′, 90 km north of narrowing observed at 78°N, with one core above the Spitsbergen continental slope and the other one above the underwater ridge.

7.4 Structure of the West Spitsbergen Current

The results presented above, both the hydrographic data and baroclinic currents calculated from the data, as well as the measured barotropic currents indicate a multi-branch structure of the Atlantic Water flow in the West Spitsbergen Current. It is of special importance for the volume and heat transport into the Arctic Ocean. Two main branches of the flow may be distinguished:

- The core of the West Spitsbergen Current.
- The western branch of the West Spitsbergen Current.

The both flows are related to the bottom topography. The main current flows along the edge of the Barents Sea and Spitsbergen shelves, with its centre above the 800 m isobath. At present, in spite of the fact that two branches of the West Spitsbergen Current were recognized, when studying the Atlantic Water transport, mostly the eastern branch of the current is taken into account. Skagseth et al. (2004) describe the flow along the continental shelf as a coherent system of barotropic currents of a length of 3,500 km, extending from the Irish-Scottish shelf to the Arctic Ocean. The authors consider the currents to be the main system, transporting heat to the Arctic Ocean. In the southern part of this current system in the NwASC, a purely barotropic is observed. In the part studied by the IOPAS we observe both barotropic flows (measured with the ADCP and LADCP) and baroclinic flows. An area of baroclinicity which induces the flow above the shelf edge is visible in each of the above-presented zonal plots. The eastern branch is a continuation of the barotropic NwASC. However, due to that the water transported along the shelf is warmer than the surrounding waters, additional baroclinic forcing is generated to enhance that current. They overlap barotropic flows, thus generating complicated flows variable in time.

The eastern branch of the WSC flows into the AO along the Spitsbergen slope as the Svalbard Branch and submerges as it gradually increases density (the AW temperature decreases), or it is rather covered by less saline (and less dense) surface waters, then it flows cyclonically along the Arctic Ocean shelf break (Rudels et al. 1999). The major part of the AW is isolated from direct contact with ice and the atmosphere by a strong cold halocline. The Svalbard Branch which flows under the halocline is the main source of the Atlantic Water for the Arctic Ocean. Having encircled the entire Arctic Ocean, the current is observed at a depth of 500 m together with other water masses leaving the AO through the FS (Coachman and Aagaard 1974). However, the AW occasionally gets in contact with the atmosphere and ice, thus releasing heat and influencing the ice cover north-east of Spitsbergen (Steel and Boyd 1998; Martinson and Steele 2001).

The western WSC branch is also steered by the bottom topography—it follows 2,000–2,500 m isobaths, however, it is a flow of another type. It is a system of baroclinic currents at the western margin of the Atlantic Domain—the Arctic Front. The fact that the flow is topographically guided indicates that it must be influenced by the bottom profile. The baroclinic currents reach the AW layer depth (down to 1,000 m), hence, also the western branch of the WSC has a barotropic component. That was confirmed by the IOPAS measurements with use of the LADCP. The western flow is much less recognized than the branch flowing along the slope. The observations suggest that the branch has much stronger tendency to meander and to shed mesoscale structures (Walczowski 1997). The process of transfrontal exchange, the AW recirculation and decay of the Arctic Front are significant for the AW transport. The transfrontal transport occurs along the entire front, however, it intensifies north of the parallel 76°N. It is certainly a process which is guided by the bottom topography (Manley 1995), however, it is also influenced by the location and the intensity of the front, and that, on the other hand, is linked to the atmospheric circulation (Schlichtholz and Goszczko 2006). The Arctic Front becomes less pronounced and bifurcates in the Fram Strait area. It is a process which takes place in stages: a portion of the AW carried by the currents flowing along the front detaches from the main flow, turns west, a part of the AW carried by the western branch even reaches the latitude of 80°N and there it recirculates southward as the Return Atlantic Current.

However, the presented zonal transects usually show more than two areas of positive (northward) flow. Also the horizontal distributions, including in particular the distribution of mean kinetic energy of flows (Fig. 6.3), show that the western branch may additionally divide at the latitude of 73°N. An intense current flowing towards north–east splits into a northward flow, along the Mohn Ridge and an eastward flow. The analysis of the bottom topography may help explain that phenomenon again. The Mohn Ridge line breaks here (Fig. 2.1), the 2,500 m isobath changes its direction from the north-eastward into the northward one, and the deep Lofoten Basin gets shallower. A portion of the flow related to the 2,500 m isobath follows this isobath northward, above the Knipovich Ridge. However, the inner portion of the flow sits over the continental slope of the Barents Sea. The phenomenon is well visible on the figure where the bathymetry is combined with mean baroclinic currents represented by mean geopotential anomaly (Fig. 7.10) which may be treated as streamlines.

The baroclinic current continuing along the northern coast of Norway and flowing into the southern part of the BSO is well visible. Flowing between the latitudes of 72°N and 74°N the current divides into two parts. The significance of topography is best visible in the area where the water flows gradually above the shallowing bottom. Depending on the angle between the streamline and the isobath, the current flows into the Barents Sea and crosses the isobaths, or diverts northward, thus following the bottom profile. Although the flows are not directly linked with isobaths but with the f/H parameter (Sect. 2.5), with such differences of latitude the sin(78)/sin(70) ratio equals 1.045, which means that the effect of change in the Coriolis parameter f between the latitude of 70 N and 78°N causes the relocation of the topographically driven current towards the depth H, larger by 4.5 % and may be omitted.

In addition to the western and eastern branch, one more central flow of the AW is observed and noted in the Fram Strait, referred to as the Yermak Branch, which name originates from the bottom structure around which it flows (the Yermak Plateau). The origin of that branch was explained splitting of the Svalbard Branch due to interaction with the bottom. In the area of study, the boundaries between individual flows are the 200 and 600 m isobaths (Manley 1995). The further route of the AW flowing into the AO from the Yermak Branch is not completely known,

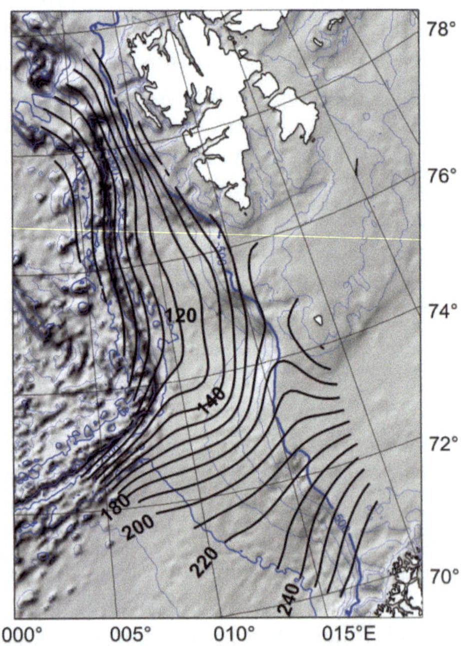

Fig. 7.10 Bathymetry (*blue lines*) and the July 2000–2007 mean geopotential anomaly at the level of 100 bar

a part of it probably surrounds the Yermak Plateau from the north and joins the Svalbard Branch, and another part recirculates with the RAC. Yet it seems that by interactions with the AO water masses, it has a stronger influence upon local conditions (including sea ice) in the Fram Strait area.

The above-presented results suggest that the source of the Yermak Branch may be the western branch of the WSC, to be more accurate, its inner warmer fraction, "confined" by the Barents Sea shelf. Having passed the arear of the narrowing of the AW flow at 78°N, the flow may enter the AO as the Yermak Branch (Fig. 7.11) Unlike the eastern or the western branch, it is not a current which is under permanent observation, however, its effects are visible south of 78°N (Fig. 5.9) and north of that area.

In the light of the above, the AW transport into the AO depends strongly upon the upstream conditions. Depending on the configuration of the western branch, it may recirculate entirely, or a portion of its water may flow into the AO as the Yermak Branch, thus increasing the heat transport into the Arctic Ocean. The processes occurring in the northern part of the WSC, west of Spitsbergen, are very significant. Due to the bottom topography, the Atlantic Domain is considerably narrowed at 78°N and the previously divided branches converge. The horizontal distributions of the AW properties (Chap. 4) clearly show that the region is a gateway which adjusts and modifies flows towards the Arctic Ocean. It is a place where the currents are characterized by sudden accelerations and the fastest flows. That is confirmed by both the LADCP measurements performed by the IOPAS and the data from the isopycnal floats.

One can state that the above diagram of currents is incoherent, since the currents near the BSO cannot flow eastward, as shown in Fig. 7.10, and northward

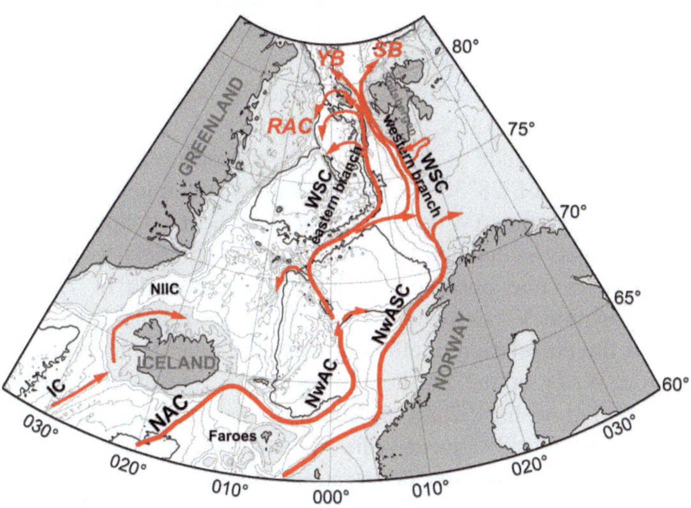

Fig. 7.11 The Atlantic Water pathways in the Fram Strait area: the Svalbard branch (*SB*), the Yermak branch (*YB*) and the Return Atlantic Current (*RAC*)

(Fig. 7.11) at the same time. However, one should bear in mind that Fig. 7.11 is only a general diagram. Sea currents are not thin lines but wide streams which may split and meander. Furthermore, the previous results showed that there are periods when the transport towards the Barents Sea or towards the Fram Strait prevails.

7.5 Signal Propagation in the West Spitsbergen Current

Knowing the structure of flow makes one want to estimate the possible velocity of signal propagation.[1] Contrary to what might seem, it is not an easy task. Polyakov et al. (2005) showed that temperature anomalies recorded at moorings in the Svinoy Section reached the Fram Strait after 18 months. That gives mean signal propagation velocity (in fact the northern V component) in the eastern WSC branch of 3.8 cm/s. Similar values were obtained as a result of comparing the data measured by the IOPAS on transect N and the published observations from the Svinoy Section (Orvik and Skagseth 2005). They described two periods of increase of the AW mean temperature by 0.6 °C: from 1997 to 1998 and from 2002 to 2003. The increase of the AW temperature in the WSC core at the latitude of 76°30′N was found by the IOPAS with a time lag of 18–21 months from the increase of temperature at 63°N. It results in 2.8–3.2 cm/s of the V component of the temperature anomaly propagation velocity. These results are very approximate since we only have summer measurements available for the transect N.

The signal propagation speed along the western branch of the WSC has not been known so far. The IOPAS work in that respect seem to be pioneer (Walczowski and Piechura 2007). In two subsequent years (2005 and 2006) baroclinic anticyclonic eddies were observed above the Mohn and Knipovich Ridges (Chap. 8). The eddies, particularly the southern one, were much larger than the mesoscale structures observed previously in the WSC western branch. In summer 2005 the eddy of 170 km diameter and the depth of the AW layer below 1,000 m was located at 73°N, whereas the second one, with a diameter of 130–160 km—at the latitude of 76°N. An eddy was observed in the Fram Strain in 2006, at the latitude of 78°30′N. The structures are much better revealed by temperature anomaly, the AW layer thickness or heat content than by absolute values. The data analysis enabled to assume that the anomaly found in 2006 in the Fram Strait was the one observed in 2005 at the latitude of 73°N. It implies the signal propagation V component along the western branch of 1.9 cm/s. Also the increase of the correlation coefficient between mean temperature and salinity of the Atlantic Water at the northern transect EB and at other transects, after applying a 1-year time lag (Chap. 5) confirms that it is possible to observe the same structures for two subsequent years. Calculations of mean advection velocity in the western part confirm the observations published by Cottier et al. 2007.

[1] a signal is considered to be a separated anomaly in water properties (temperature and salinity).

In February 2006 a sudden increase of water temperature in Kongsfjord (a fjord in the north-west Spitsbergen) was observed. It was probably passing of the northern anomaly visible in Fig. 8.4. It results in a similar propagation velocity of approximately 2 cm/s and confirms that the anomaly observed in summer 2006 was the southern one from 2005.

The measurements show that there is a difference in the signal propagation in the eastern and western WSC branch. The signals also have different structures—the increase of temperature in the western branch may be caused by both the general increase in water temperature and the passage of a large anticyclonic eddy. In the eastern part the increase in temperature is related to the AW upstream temperature rise. The heat content and transport by both branches evolves in time.

The presented results show that the western branch may also have significant impact on the warming observed in the Fram Strait and the Arctic Ocean (Walczowski and Piechura 2006, 2007). Currents in that branch are characterized by higher variability than the currents in the WSC core. In 2006 and 2007 there was a significant increase of heat transport by the western branch. It is worth noting that the both branches constitute the boundaries of the Atlantic Domain, there is a thick layer of the Atlantic Water in between them of heat content and salinity higher than at the edges. Water in the Domain centre does not mix with colder and less saline Polar and Arctic waters as easily, thus it maintains the properties (salinity in particular) characteristic for the source and may carry them farther north. Temperature and heat content of the AW layer is modified here mainly by the exchange with the atmosphere and vertical mixing, including in particular the winter convection. However, the highest temperature (not heat content, though) is found in the barotropic WSC core.

7.6 Volume and Heat Budget

The obtained averaged results force to make an attempt to develop a total budget of volume and heat transported across the transects. The main problem is proper determination of the currents velocities. Despite the progress in technology, measuring currents over large distances and a long periods is still a complicated and costly task. That is why the analysis is limited to baroclinic calculations. It provides an indirect possibility to estimate the budget of heat supplied by the thermohaline circulation. The question concerning the energy fluxes through the ocean surface in summer is not considered here. Based on the data from the R/V Oceania, such considerations were undertaken by Jankowski (1991), Druet (1993), and an accurate description of the components of the energy budget was provided by Piechura et al. (2002). The present paper provides only a bulk estimate of the AW volume and heat budget, which assumes that, due to slow advection, changes in the water properties observed in the northern part, could occur much earlier, thus, the components of the budget may be close to averaged annual values.

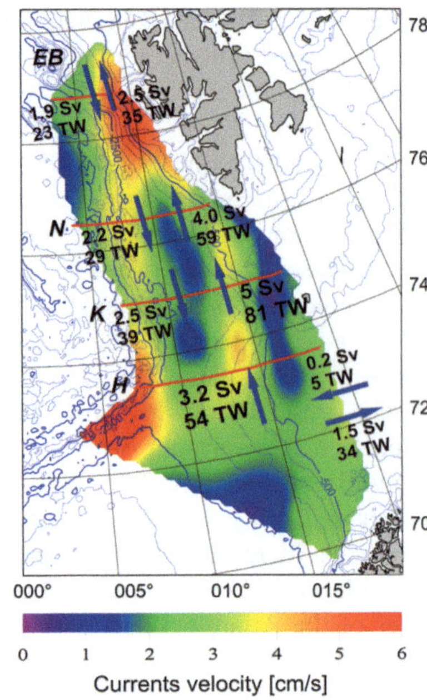

Fig. 7.12 The July 2000–2007 mean current velocity (*colour scale*) and baroclinic volume and heat transports across the studied transects

Figure 7.12 shows the values of mean baroclinic heat and volume transport in both directions across the above transects. In the case of transect H, due to the above-mentioned reasons, a decision was made to show only the flux obtained from extrapolation of data from the downstream transects. Next, mean AW properties from 2000 to 2007, measured and calculated at four zonal transects, were gathered in Table 7.3. The data for the areas between the transects were also collected. Based on the relationships between the AW transport and properties, mean advection velocities, heat fluxes to the atmosphere and the exchange with adjacent sea regions, mostly through the Arctic Front, were estimated. Approximate values were presented, since the only a rough estimate of the budget is possible with available data. However, even such rough estimates are very valuable, because the range of volume and heat transport values in the Arctic is very wide. The table includes a description how further results were obtained.

- On the basis of the AW area at transects (5) and the AW transport across the transects (7), mean velocities required to obtain that transport (8) were calculated;
- On the basis of heat content at transects (6), mean velocities required to obtained given heat transport (10) were calculated;
- Further calculations were made with use of velocities (10) which give mean velocity of 1.6 cm/s, a bit lower than the velocity of signal propagation in the western branch, as previously estimated in the paper;

Table 7.3 Basic AW properties on the investigated transects and elements of the balance of heat and volume

No.	Transect properties	H	K	N	EB
1	Latitude	**73°30**	**75°00**	**76°30**	**78°50′**
2	Mean AW temperature (°C)	3.94	3.36	3.04	2.69
3	Mean AW salinity (psu)	35.064	35.046	35.033	35.013
5	The AW area at transect (km^2)	211	163	142	82
6	Heat content at 1 m of transect (10^{15} J/m)	3.5	2.3	1.8	1.0
7	Resultant baroclinic volume transport of the AW (Sv)	**3.2**	**2.5**	**1.8**	**0.6**
8	Mean velocity necessary to provide volume transport (cm/s)	1.5	1.5	1.3	0.7
9	Resultant baroclinic heat transport (TW)	**54**	**42**	**30**	**12**
10	Mean velocity necessary to provide heat transport (cm/s)	1.5	1.8	1.7	1.2
11	Northward volume transport (Sv)	3.2	5	4	2.5
12	Northward heat transport (TW)	54	81	59	35
13	Heat/volume transport (TW/Sv)	16.9	16.2	14.7	14
14	The necessary AW temperature to provide heat transport (°C)	4.1	3.9	3.7	3.5
	Properties between transects		**H–K**	**K–N**	**N–EB**
15	The Atlantic domain area between transects (km^2)		51,852	47,287	50,954
16	Distance between transects (km)		166	166	255
17	Time (days) for transport between transects with current necessary for the heat transport		120	120	184
18	Transport of the AW across the Arctic Front (Sv)		**0.7**	**0.7**	**1.2**
19	Transport of the AW across the AF per 100 km (Sv/100 km)		0.42	0.42	0.48
20	Mean heat content in the AW column (GJ/m^2)		8.0	7.3	5.6
21	Total heat content in the AW layer (10^{20} J)		4.15	3.43	2.86
	Calculated mean values in areas between transects			**HK–KN**	**KN–NEB**
22	Differences in heat content in the areas between transects (GJ/m^2)			0.7	1.7
23	Mean time for advection between the areas (days)			120	152
24	Heat flux to the atmosphere (W/m^2)			**67**	**129**
	Values calculated for the areas		**H–K**	**K–N**	**N–EB**
25	Heat flux to the atmosphere (TW)		**3.5**	**3.2**	**6.6**
26	Heat transport across the AF (TW)		**8.5**	**8.8**	**11.4**
27	Heat transport across the AF per 100 km (TW/100 km)		5.12	5.3	4.47
28	The AW temperature necessary to transport heat across the AF with a given volume transport (°C)		3.0	3.1	2.4
	Calculated mixing coefficients				
29	Coefficient α		0.968	0.946	0.962
30	ΔT value		−0.49	−0.19	−0.27

- Mean AW temperature required to obtain given heat transport (14) was calculated from the northward volume transport (11) and the northward heat transport (12);
- The AW transport across the Arctic Front (18) was calculated from the difference in the volume transport across individual transects;
- Next, differences of mean values for the entire areas were used;
- Mean heat fluxes to the atmosphere for given regions (24) were calculated from differences of heat content in the AW column in individual areas (22) and mean time of advection (23);
- The values of fluxes were used as mean values for calculating transports (25);
- Heat transport across the Arctic Front (26) was calculated from the budget;
- Knowing the area between transects and heat fluxes, heat transport to the atmosphere in individual areas was calculated (27);
- In order to verify the results, the temperature of the AW required to obtain given heat transport heat transport across the AF (28) was calculated (with the previously known volume transport);

However, changes in the AW temperature observed in the north result not only from the heat exchange with atmosphere but also from mixing. Therefore, the heat fluxes to the atmosphere were corrected, taking into consideration mixing with colder surrounding waters. The simplest model of mixing takes into account 3 water masses: the AW in the south (AW_S), the AW in the north (AW_N) and the Polar Water (PW). Changes in the AW temperature and salinity were taken into account.

$$\begin{aligned} SAW_N &= \alpha * SAW_S + (1 - \alpha)SPW \\ TAW_N &= \alpha * TAW_S + (1 - \alpha)TPW + \Delta T \end{aligned} \tag{7.3}$$

The unknown α is the fracion of the AW in the final water mass, ΔT is the change of the AW temperature resulting from heat flux to the atmosphere. The following properties of the Polar Water were assumed: temperature = 1 °C, salinity = 34.5. The following values were obtained for temperature and salinity at transects H and EB listed in Table 7.3: $\alpha = 0.91$ $\Delta T = -0.98$ °C.

It means that the Atlantic Water at the transect EB contains 91 % of the Atlantic Water, originating from the transect H. In the decrease of the AW mean temperature between these from 3.94 to 2.69 °C was caused by exchange with the atmosphere. Further decrease by which is 22 % of total AW cooling between transect H and EB resulted from mixing with surrounding waters.

Partial calculations for individual regions have values as given in (29) and (30). Similarly to other cases, approximately 20 % of the AW temperature decrease was caused by mixing with surrounding waters. However, that does not influence heat flux to the atmosphere, since it is defined on the basis of heat content in the AW layer and it does not change as a result of mixing.

As a result of those calculations, heat and volume budget for the region of study was established (Fig. 7.13). It includes the part north of the parallel 73°30′N and

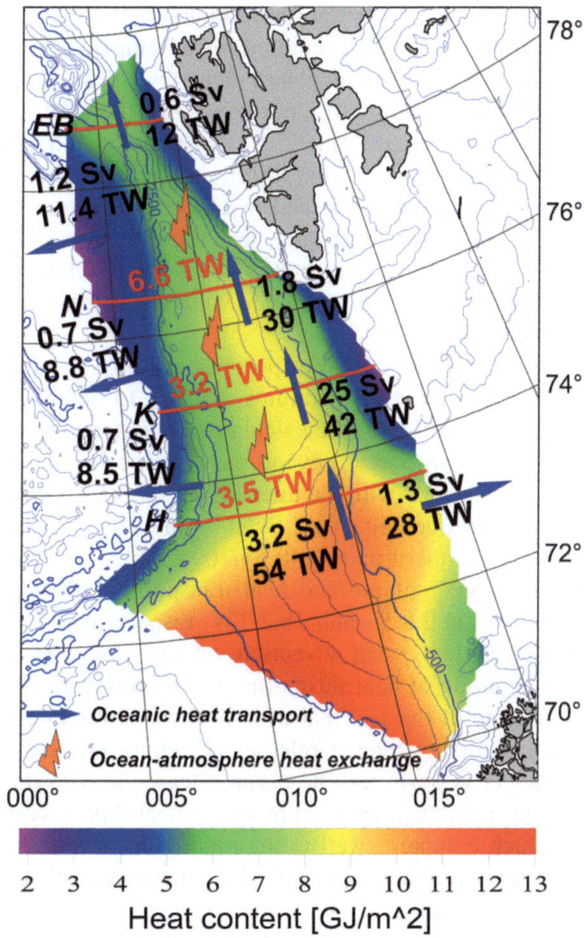

Fig. 7.13 Balance of heat and volume transported by the baroclinic currents. Transects from 7.3 are marked

transport to the Barents Sea across transect V1. Similarly to Table 7.3, the transport across the main WSC stream is denoted in the table as a transport across the Arctic Front. It also includes eastward transport, since the AW transport through the Storfjordroenna, between the Bear Island and Soerkapp, cannot be excluded. The IOPAS observations indicate a significant recirculation of the AW in that area.

The analysis of the results presented in Table 7.3 and in Figs. 7.12 and 7.13 leads to some conclusions and comments:

- The estimated mean baroclinic volume and heat into the Barents Sea is 1.3 Sv and 29 TW, respectively. The volume transport is close to the specified value of 1.8 Sv (Skagseth 2008), the heat transport is much lower than their value of 48 TW.

- 22.3 TW of heat is transported into the Barents Sea per 1 Sv of the AW volume transport. It gives a required mean temperature of the AW of 5.4 °C, which is very close to the calculated mean (Fig. 4.2).
- In the Fram Strait, at the transect EB, the northward baroclinic transport of the AW was estimated on 2.5 Sv as compared to 5.8 Sv of the northwards transport given by Schauer et al. (2008). Furthermore, the AW definition used by Schauer was different and resulted in a lower transport;
- The baroclinic northward heat transport across transect EB was estimated on 34 TW as compared to 40 TW given by Schauer et al. (2008);
- The calculated net baroclinic volume and heat transport are 0.6 Sv and 12 TW, respectively.
- The northward heat transport per a unit of volume decreases with latitude (13);
- The calculated temperatures required to obtain given heat transports (14) correspond well with the calculated mean temperatures of the AW column;
- Heat is transported beyond the Atlantic Domain—mostly westward across the Arctic Front—with the intensity of 9.5–12.1 TW/Sv. The volume flux maintains similar values—ca. 0.42 Sv/100 km—along the entire front, and it slightly rises in the northern part (0.48 Sv/100 km);
- Total AW volume and heat transports across the Arctic Front in the studied region are 2.8 Sv and 28.7 TW, respectively;
- The applied mean advection velocity does not affect the budget but only the magnitude of heat fluxes (not the total transport) to the atmosphere;
- In spite of the fact that even at those latitudes ocean absorbs in summer much more energy than it releases (Piechura et al. 2002), the total calculated heat budget is positive—the ocean releases heat to the atmosphere;
- The calculated ocean–atmosphere heat flux is within the range of 67–129 W/m^2, which compares well to the more general annual value of 100 W/m^2 given by Furevik and Nilsen (2005). It is also correct that the value of the flux increases with latitude.
- The results indicate that the baroclinic currents alone, being the main mechanism of thermohaline circulation, are capable of supplying enough Atlantic Water to support the ocean–atmosphere heat fluxes as described in various references.

The last conclusion is particularly meaningful—it confirms the role of the thermohaline circulation for the climate. It also explains the previously presented correlations between mean annual temperature in Hornsund and the AW temperature. The AW transport across the Arctic Front plays a significant role. The AW is not only the main component of deep water formed in the Greenland Sea, it also influences melting of sea ice carried from the Arctic Ocean by the East Greenland Current. The recent retreat in 2007 indicates that the mechanism may be very important—by melting ice along the Greenland shelf, the AW allows for larger advection of sea ice from the AO.

7.7 Summary

Chapter 7 was focused on presenting the structure of sea currents in the studied region, the baroclinic transport of volume and heat across individual transects, the variability of fluxes in space (meridionally) and time. Existence of a negative correlation between the baroclinic currents into the Barents Sea and towards the Fram Strait is a valuable observation.

Next, the structure of baroclinic and barotropic currents was compared. Emphasis was put on the fact that there are at least two branches of the WSC which, steered by the bottom topography, approach each other to form a complex system of currents in the Fram Strait. Furthermore, a suggestion was made that a central stream (the Yermak Branch) is formed by splitting of the western branch in the southern part of the study area. Concepts and estimations of the mean velocity of signal propagation in both branches of the WSC were presented.

Finally, an attempt to calculate a simplified volume and heat budget was made. Although the results represent only baroclinic transports, they provide realistic results. Proving that the baroclinic currents alone supply significant amount of heat to the Arctic climate is an important achievement.

References

Coachman LK, Aagaard K (1974) Physical oceanography of arctic and subarctic seas. In: Herman Y (ed) Marine geology and oceanography of the Arctic Seas. Springer, New York, pp 1–81

Cottier FR, Nilsen F, Inall ME, Gerland S, Tverberg V, Svendsen H (2007) Wintertime warming of an Arctic shelf in response to large-scale atmospheric circulation. Geophys Res Lett 34:L10607. doi:10.1029/2007GL029948

Druet C (ed) (1993) Polar Marine Research (2), The results of Polis oceanographic investigations focused on interannual variability of the Greenland Sea energoactive zones, Stud. i Mater. Oceanol., 65, p 222

Furevik T, Nilsen J (2005) Large-scale atmospheric circulation variability and its impacts on the Nordic Seas ocean climate—a Review, [in:] The Nordic Seas: An integrated perspective, AGU Geophys Monogr Ser 158, 105–137

Jankowski A (1991) Heat flux and wind momentum flux evaluation in certain regions of the Norwegian Sea in summer 1987 (based on results of the AREX-87 cruise), Studia i Materiały Oceanologiczne KBM PAN. Polar Marine Res 1(58):77–89

Manley TO (1995) Branching of Atlantic water within the Greenland-Spitsbergen passage: an estimate of recirculation. J Geophys Res 100(C10):20627–20634

Martinson DG, Steele M (2001) Future of the Arctic sea ice cover: implications of an Antarctic analog. Geophys Res Lett 28(2):307–310

Maslowski W, Newton B, Schlosser P, Semtner A, Martinson D (2000) Modeling recent climate variability in the Arctic. Geophys Res Lett 27, 3743–3746

Orvik KA, Skagseth Ø (2005) Heat flux variations in the eastern Norwegian Atlantic Current toward the Arctic from moored instruments, 1995–2005. Geophys Res Lett 32:L14610. doi:10.1029/2005GL023487

Osinski R, Wieczorek P, Beszczynska-Moller A, Goszczko I (2003) ADCP-referenced geostrophic velocity and transport in the West Spitsbergen Current. Oceanologia 45(3):425–435

Piechura J, Beszczynska-Moeller A, Osinski R (2001) Volume, heat and salt transport by the West Spitsbergen Current. Polar Res 20(2):233–240

Piechura J, Osiński R, Petelski T, Woźniak SB (2002) Heat and salt fluxes in the West Spitsbergen Current in summer time. Oceanologia 443:307–321

Polyakov IV et al (2005) One more step toward a warmer Arctic. Geophys Res Lett 32:L17605. doi:10.1029/2005GL023740

Rudels B, Fredrich HJ, Quatfasel D (1999) The Arctic circumpolar boundary current. Deep-Sea Res II 46:1023–1062

Schauer U, Beszczynska–Moeller A, Walczowski W, Fahrbach E, Piechura J, Hansen E (2008) Variation of measured heat flow through the fram strait between 1997 and 2006, Arctic-Subarctic Ocean fluxes. Springer Science, pp 15–43

Schlichtholz P, Goszczko I (2006) Interannual variability of the Atlantic water layer in the West Spitsbergen Current at 76.5°N in summer 1991–2003. Deep-Sea Res I 53, doi:10.1016/j

Skagseth Ø (2008) Recirculation of Atlantic water in the western Barents Sea. Geophys Res Lett 35:L11606. doi:10.1029/2008GL033785

Skagseth Ø, Orvik KA, Furevik T (2004) Coherent variability of the Norwegian Atlantic Slope Current derived from TOPEX/ERS altimeter data. Geophys Res Lett 31:L14304. doi:10.1029/2004GL020057

Steele M, Boyd T (1998) Retreat of the cold halocline layer in the Arctic Ocean. J Geophys Res 103:10419–10435

Walczowski W (1997) Transfrontalna wymiana masy i ciepła w rejonie Frontu Arktycznego, PhD Thesis, IOPAN, p 123

Walczowski W, Piechura J (2006) New evidence of warming propagating toward the Arctic Ocean. Geophys Res Lett 33:L12601. doi:10.1029/2006GL025872

Walczowski W, Piechura J (2007) Pathways of the Greenland Sea warming. Geophys Res Lett 34:L10608. doi:10.1029/2007GL029974

Walczowski W, Piechura J, Osinski R, Wieczorek P (2005) The West Spitsbergen Current volume and heat transport from synoptic observations in summer. Deep-Sea Research I 52:1374–1931

Chapter 8
Changes in the Ocean Climate

Even very slow mean currents of ca. 1.6 cm/s velocity are of particular significance for advective transport of heat, since they influence a local climate, and due to the specific nature of the studied area, also the global climate. It was described in Chap. 7. In situ observed currents are more intense and variable. It results in oceanic transport and ocean–atmosphere heat fluxes larger than described. We have witnessed important climate change in recent years.

8.1 The Northward Propagation of Warm Water

The Atlantic Water properties in the Nordic Seas changed a lot in the period described in the paper (summer 2000–summer 2007). The IOPAS oceanographers observed an increase of the AW temperature in 2004, record values in 2006 and then a decrease in 2007. It was connected with changes in salinity.

Since the measurements were taken only once a year, some doubts could arise whether the observed variability represents the inter-annual climate changes or just some shifts in the seasonal cycle. Calculations based on the results of a numerical model show that in the case of the WSC it is possible to determine a multi-year trend based on data collected only during the summers (Chap. 4). Also the analysis of salinity which undergoes much smaller seasonal changes than temperature, and, concurrently, is strongly correlated with temperature, shows that the observed anomaly is in fact due to the increase of the Atlantic Water temperature. Integrated values, such as heat content, and average values, such as mean temperature of the entire AW volume in the AREX study area, were presented earlier. High correlation between mean temperature of the entire AW mass in the AREX study area and the mean annual temperature of air in Hornsund is a meaningful result. Chapter 9 will present some additional aspects of the observed rise in the AW temperature.

In the period of 2004–2006 a northward transport of warm, highly saline water was observed (Fig. 8.1). The AW properties changed significantly in the area of the western Spitsbergen. Warm Atlantic-origin water flowed onto the shelf and into the fjords of the western Spitsbergen. Changes of the AW properties were observed

W. Walczowski, *Atlantic Water in the Nordic Seas*,
GeoPlanet: Earth and Planetary Sciences, DOI: 10.1007/978-3-319-01279-7_8,

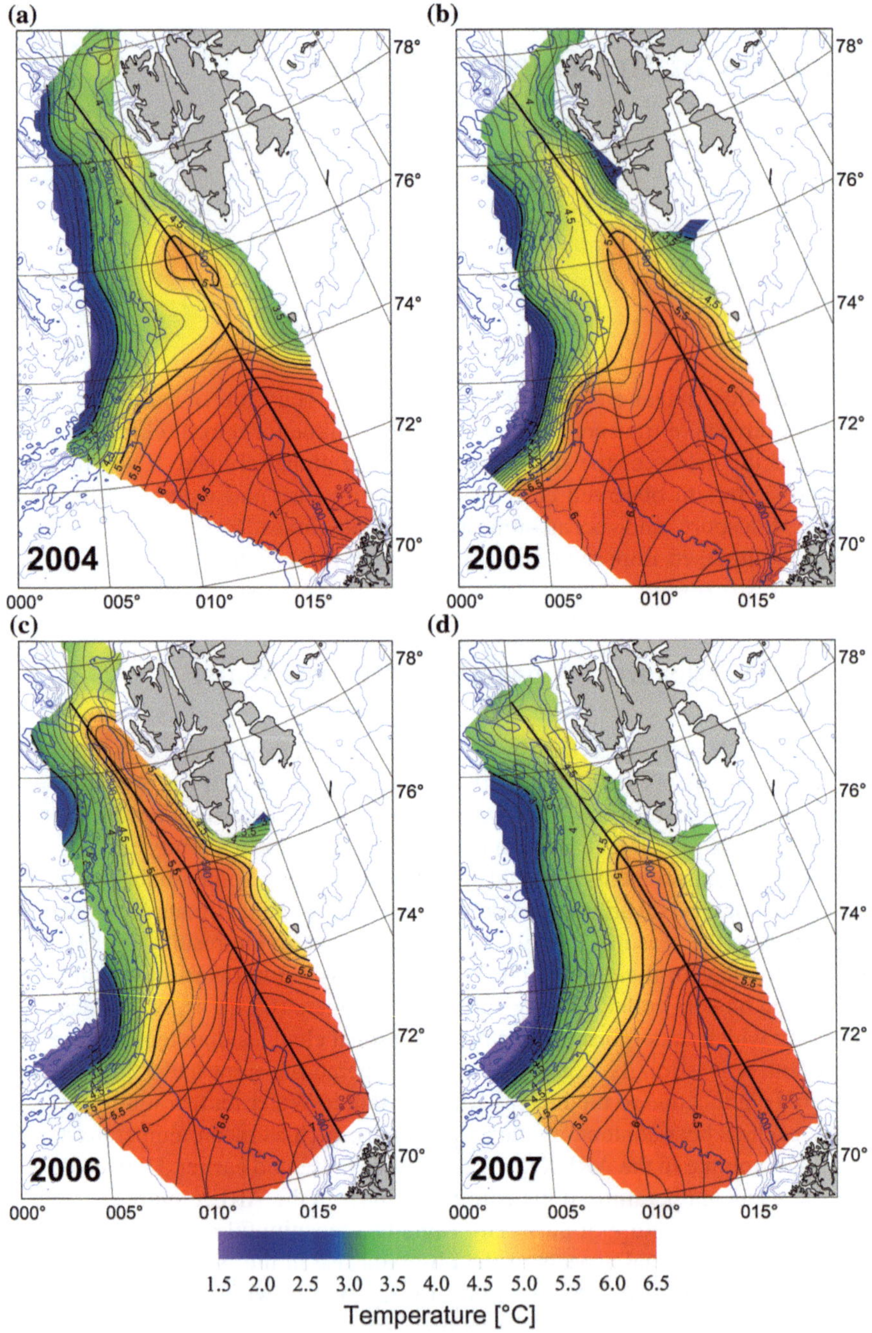

Fig. 8.1 Evolution of the AW temperature at a level of 100 m in the period 2004–2007

both in mean values (Chap. 4) and at specific levels. In 2004 the 5 °C isotherm at a depth of 100 m reached 74°30′N, the area of slightly increased temperature extended farther north. In 2005 the 5 °C isotherm passed the latitude of 76°30 N, and in summer 2006 it reached 79°N(Fig. 8.1). The location of this isotherm is considered here as an indicator of warming or the ocean climate change.

For the advective flow a local increase of temperature is caused by a movement of warmer waters, in the case of the WSC from the south to the north. However, as shown previously, the AW undergoes a significant modification during the advection. Total change of temperature in a given point includes an advective component and a local component:

$$\frac{dT}{dt} = u\frac{\partial T}{\partial x} + \frac{\partial T}{\partial t} \tag{8.1}$$

A change in temperature at any point may be caused by one of the following three factors, or a combination of them:

- a change in the advection velocity u;
- a change in the horizontal temperature gradient $\frac{\partial T}{\partial x}$
- a change in the local component $\frac{\partial T}{\partial t}$—in that case it is a temperature change due to ocean-atmosphere fluxes, or mixing with the surrounding waters.

In order to examine the changes, it is necessary to know the horizontal temperature gradients and mean advection velocity.

For that purpose a line was selected (indicated in Fig. 8.1) which marks the axis of the warm water tongue, along which temperature changes in individual years and its gradients at the level of 100 m were calculated (Fig. 8.2). One line was chosen for all cases, to provide better adjustment of the bottom profile, breaking at the latitude of 76°N. It is worth noting that in 2005 the area of maximum of gradients in warm water tongue (the WSC axis) was shifted away from the determined line.

Between the latitudes of 71°N and 73°N mean horizontal temperature gradient $\frac{\partial T}{\partial x}$ in July 2003–2007 varied between 0.1 and 0.25 °C/100 km. It is characteristic that along the given line, the AW temperature at 72°N is very much similar in all years, changing between 6.4 °C in 2003 and 6.8 °C in 2006. Only farther north, near the latitude of 74° temperatures vary significantly. At the latitude of 74°N and 76°N the temperature difference between the coldest and the warmest year was 1 °C, and at 78°N it rises to 1.8 °C. In the Fram Strait, at the latitude of 79°30′N the difference of temperatures falls to 0.4 °C again. Attention is drawn by a step-like character of temperature lines. The first sudden decrease of temperature is observed in the area between 73°N and 74°N—the Barents Sea Opening. The temperature changes here probably as a result of mixing with the AW recirculating in the Bear Island Trough. It is confirmed by the analysis of salinity variability. Up to the latitude of 72°30′N where maximum salinity is found, the impact of less saline Norwegian Coastal Current is clearly visible. The decrease of salinity north of 72°30′N is correlated with the decrease of temperature, which indicates that the

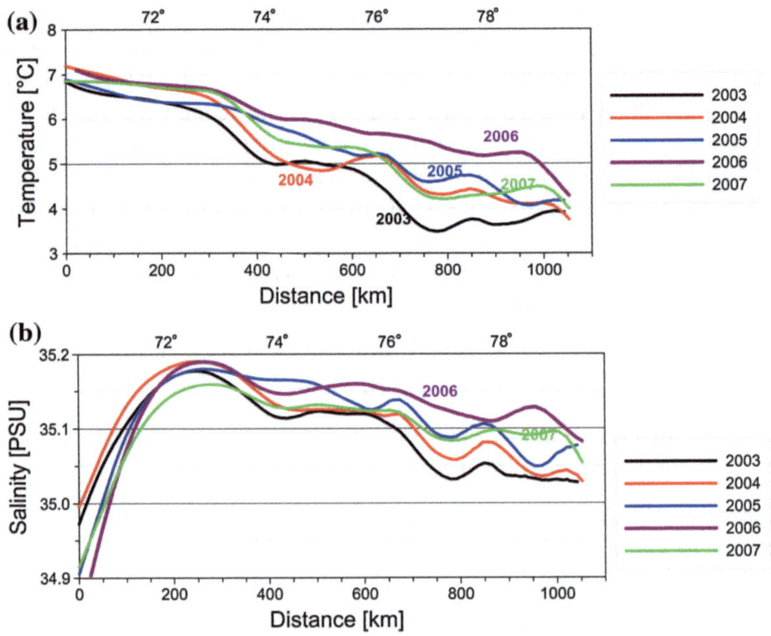

Fig. 8.2 Temperature (**a**) and salinity (**b**) of the AW at the level of 100 m, along the WSC axis, obtained from the interpolated data

AW transformation is due to mixing with colder, less saline waters of the Barents Sea. The next "step" visible on the temperature and salinity plots is the front of the warmest AW fraction tongue, the area of strong salinity and temperature gradients. It is the area which shifted 300 km up north at a velocity of 0.9 cm/s in the period 2004–2006. It is interesting that from 2004 to 2006 the step-like structure gradually reduced and horizontal temperature gradients were getting lower. It also concerns the salinity lines. In 2006 the temperature and salinity lines reached the lowest inclination.

The 2004–2006 mean horizontal temperature gradient $\frac{\partial T}{\partial x}$ for the 700 km section between the 250th and 950th km of the investigated transect was 0.3 °C/100 km. Mean advection velocity u in the WSC core was estimated on 3 cm/s. According to Eq. 8.1, if there was no heat exchange surrounding water masses and atmosphere ($\frac{\partial T}{\partial t} = 0$), the AW temperature in the WSC core would rise due to advection by 2.8 °C/year. If AW temperature does not rise, there is equilibrium between the inflow of warm AW and local cooling by releasing heat to the atmosphere and mixing with cooler waters. For the purpose of simplification the second process will not be considered at the moment. Cooling of a 200 m water column by 2.8 °C results in 76 W/m^2 heat flux to the atmosphere per year. That value is in the lower range of fluxes for the given region, however, one should bear in mind that the AW releases heat from a layer thicker than 200 m.

During the period of study temperature rose by 0.6 °C/year on average and advective increase of temperature by that value requires increasing average velocity of currents by 0.7 cm/s or increasing horizontal temperature gradients by 0.06 °C/100 km. In the period 2004–2006 no increase of temperature gradient along the WSC axis was noted. On the contrary, in 2006 the gradient decreased significantly. Thus, the cause of the temperature increase lied in a change of local conditions of exchange or in acceleration of the flow. The results presented in Chap. 6 suggest that the main cause of the observed change was the increase of currents velocity. The V_{30} component value in the northern part of the study area rose by 2 cm/s between summer 2004 and summer 2005.

It seems that the observed shift of the warm water extent may occur quite easily, however, the feature is not frequent. Isotherms were shifted northwards for two subsequent years: in 2005 and 2006. In summer 2007 situation returned to that from 2005. Also the mean value of the V_{30} component in the northern area of study decreased. Therefore, it can be argued that one of the most important causes of the warming observed in the northern part was the increase of the WSC velocity in that region, and that in 2007 the system returned to a state of dynamic equilibrium. The increase of the western branch activity was also a significant cause of the warming.

8.2 Changes in the Properties of the Atlantic Water in the Eastern and Western Branches

When studying the AW warming in the Fram Strait area, it is also meaningful to determine the heat transport process. The WSC does not represent a single stationary flow of warm Atlantic Water, it is rather a complicated system of currents driven by various forcings. In general, these currents carry the Atlantic Water transported by the North Atlantic Current, however the water has different physical properties. The eastern branch is a barotropic-baroclinic flow connected with the edge of continental shelf, whereas the western branch is mainly a baroclinic current flowing along the system of underwater ridges. The fact that they are independent flows of different dynamics is well visible in time series of temperature in the western and the eastern parts of transects. The difference of temperature between the western and the eastern flow at the southern transect H (73°30′N) is the largest, it decreases only in 2005 due to the activity of mesoscale structures in the western branch. The farther north, the lower the difference in temperature in both branches and the AW properties become similar (Fig. 8.3) when both flows cool down and mix.

It is characteristic that the WSC core temperature (the eastern part) is almost constant in time at the transect H. Only farther north high variability of that value is more distinct. According to the previously presented results, it must be related to a change in exchanges through the BSO (intensification of the baroclinic transport into the Barents Sea is correlated with the decrease of the WSC temperature the following year), as well as with a change in intensity of the AW recirculation in the Bear Island Trough. Also the mesoscale activity must affect local changes in

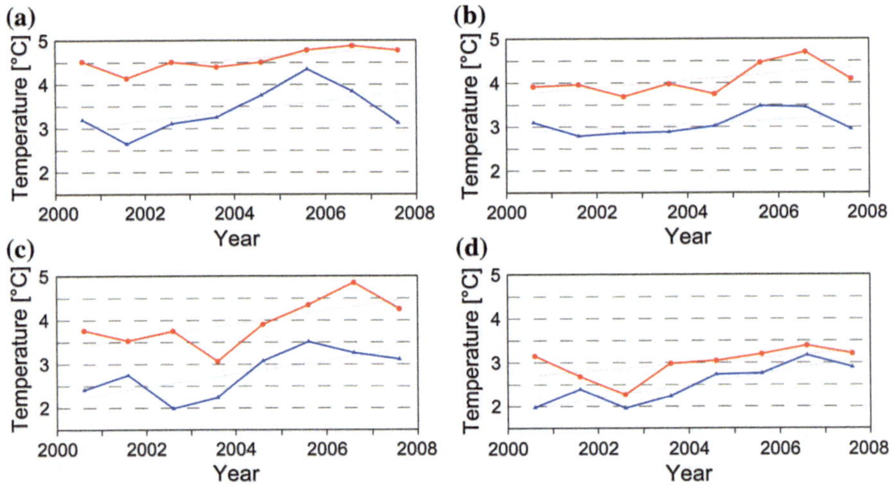

Fig. 8.3 Time series of temperature in the western (*blue line*) and the eastern branch (*red line*) of the WSC on transects H (**a**), K (**b**), N (**c**) and EB (**d**)

Table 8.1 Correlation coefficients for temperature–temperature and salinity–salinity in the eastern and western branch of the WSC at sections

Property	H	K	N	EB
Temperature	0.54	**0.79**	0.52	0.50
Salinity	**0.93**	**0.90**	0.54	0.56

temperature. In the case of the western branch, the mesoscale structures influence its temperature along its entire length. The impact was particularly high in 2005 and 2006 when propagation of great anticyclonic eddies were observed. At the horizontal distribution of temperature from 2005 (Fig. 8.1), the impact of the western branch upon the field of temperatures between the latitudes of 72°N and 74°N is clearly visible.

The correlation coefficients between AW salinity in the eastern and western branches are bigger than between temperature (Table 8.1). Both correlations decrease with latitude.

8.3 The Importance of Mesoscale Eddies in Heat Transport

The structure of heat transport and its recent variability are well pictured by the anomalies of heat content in the AW layer (Fig. 8.4). Vectors of baroclinic currents anomalies have been overlaid on colour map of heat content anomalies. In cold years, the heat anomalies are circular in shape—they are baroclinic eddies of

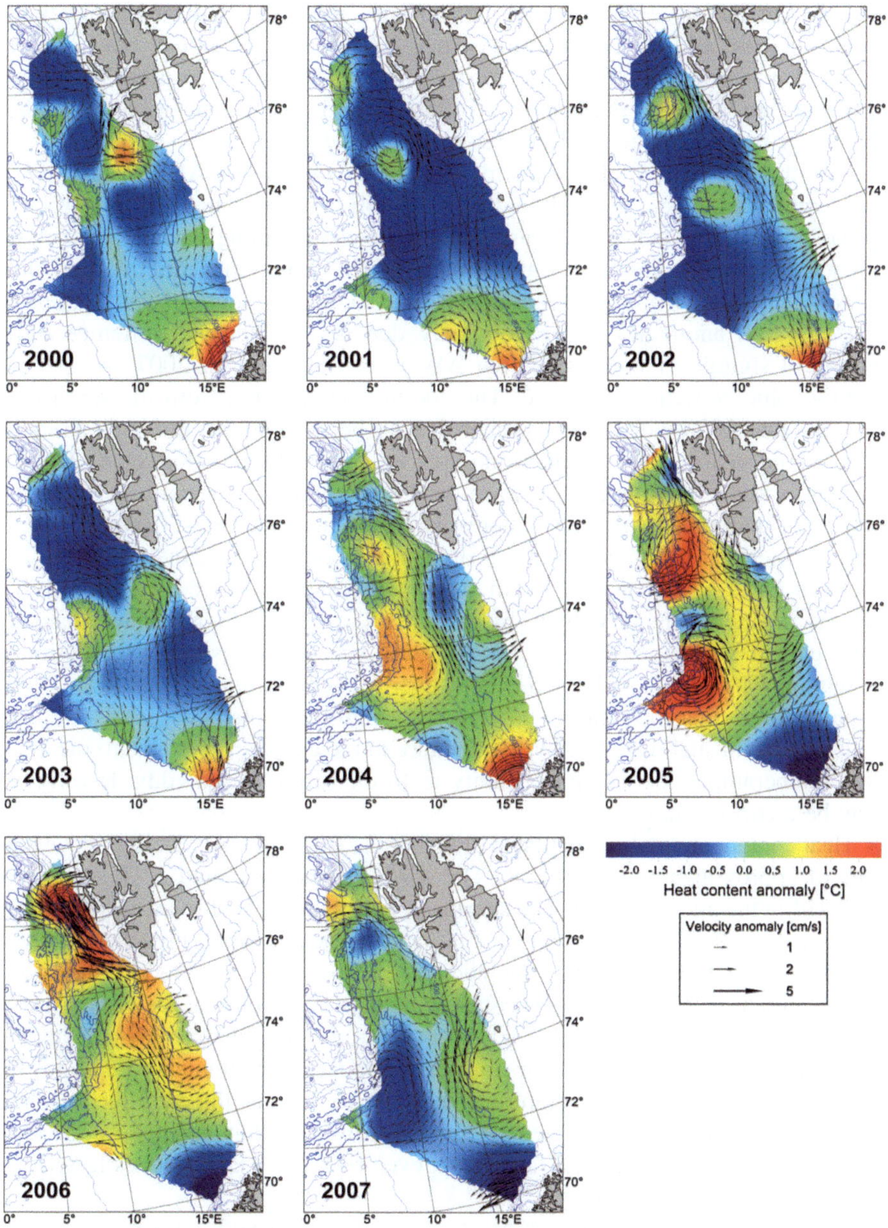

Fig. 8.4 Heat content anomalies in the Atlantic Water layer (GJ/m^2)—colour scale, and anomalies of baroclinic currents velocities at the level of 100 dbar (vectors). Anomalies calculated against the July 2000–2007 mean

ca. 100 km diameter. The baroclinic currents anomalies indicate anticyclonic nature of those eddies. The warming since 2004 has larger extent, but it clearly shows that it originates from the western branch. In 2005 two great mesoscale structures are visible at the boundary of the Arctic Front. Diameters of the eddies are nearly twice as large as structures observed before.

Also the AW layer thickness anomalies (not shown) and the ADCP and LADCP measurements indicate the eddy nature of the observed structures. The AW layer in those eddies is thicker and warmer. In 2006 the warming occupied the entire domain and the maximum of heat anomaly shifted in the Fram Strait region. Heat content increased also in the eastern part of the study area, south of Spitsbergen. In that case the anomalies occurred mostly due to the increase of temperature, without a significant increase of the AW layer thickness. In 2007 the average hydrographic conditions returned. The cooling approaching from the south-west suggests that 2008 will not be warmer, it may be even colder. It is symptomatic that in cold years heat content in the area influenced by the Norwegian Coastal Current is above the average, in warm years the anomaly is usually negative.

2005 is undoubtedly the most interesting of the analyzed years. Great baroclinic eddies which transfer huge quantities of heat were observed for the first time. The southern eddy was at least 170 km in diameter and it lowered the level of all isolines (deepened the AW layer) by at least 250 m (Fig. 8.5). Minimum diameter and depth are specified here, since the eddy was not measured with optimum resolution. The eddy could reach even deeper between station H14 and H13, there is no certainty that the transect crossed it through the centre.

It is also unusual that the AW temperature in the eddy is very close to the WSC core temperature. There is no certainty as to the origin of the eddy, however, it may be a sort of clue.

Heat content anomaly (calculated along the 2 GJ/m^2 anomaly isoline) covers an area of approximately 16,000 km^2. The AW layer accumulated 1.94×10^{19}J of heat within the anomaly area. The surplus of heat as compared to mean value from the period 2000–2007 for the given area (positive heat anomaly) reached as much as 5.5×10^{19} (5,500 PJ). Thus, the eddy is an extremely effective in transporting the oceanic heat. The structure is coherent, thanks to its rotation the eddy is much more stable and robust than a baroclinic along-front flow. That is why an eddy has a long lifetime and is capable of transferring most of accumulated heat into the Arctic Ocean. The AW temperature changes on its way, as the eddy releases heat to the atmosphere. However, the heat anomaly, always calculated against the local heat content remains the same, and it can even rise, if the eddy moves faster than an average advection velocity.

Observations made in the Fram Strait area in 2005 and 2006 provide some information whether such eddies are capable to reach the Arctic Ocean. In 2005 the northern heat anomaly also manifested as an anticyclonic eddy with its centre above an underwater ridge. Figure 8.4 shows that the eddy is in a shape of ellipsoid, the flow is disturbed due to interactions with the bottom in the previously described location where isobaths converge at the latitude of 78°N. Distribution of the velocity anomaly shows that in relation to an average flow, the eddy will shift

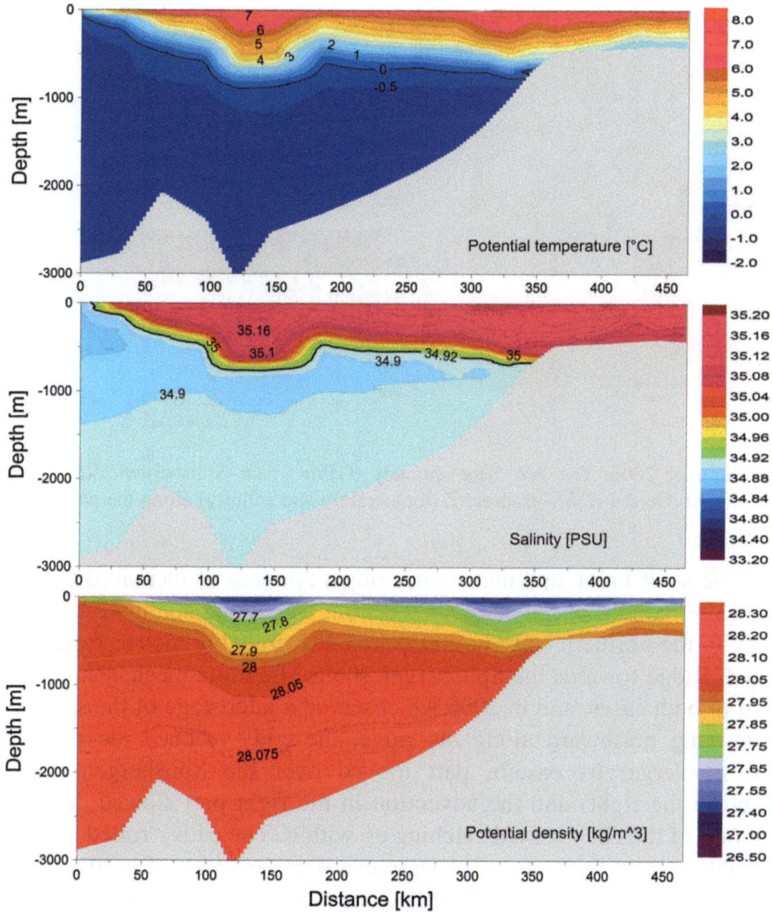

Fig. 8.5 Temperature, salinity and potential density on transect K in 2005

towards the north-east, above the Spitsbergen shelf break, to merge with the Svalbard Branch which flows into the Arctic Ocean. The northern anomaly is smaller than the southern one. It is confined by the 2 GJ/m^2 isoline, it covers the area of 6,700 km^2 and contains 2.3 × 10^{19} J of heat surplus. If the boundary is shifted to the 1.5 GJ/m^2 isoline, the anomaly will cover the area of 21,000 km^2 and contain 4.2 × 10^{19} J of heat surplus as compared to the average heat content of the AW layer in that region. The author (Walczowski and Piechura 2007) claims that the anomaly was the cause of sudden increase of water temperature in 2006 in Kongsfjord (Cottier et al. 2007). The thesis stating that such an eddy may reach the AO after transformation is confirmed by measurements performed in 2006. The anomaly observed west of Spitsbergen was large, it stretched zonally over more than 300 km, however, its core had a form of ellipsoidal eddy of 130–160 km diameter (Fig. 8.6). The eddy centre was located in the narrowing flow area

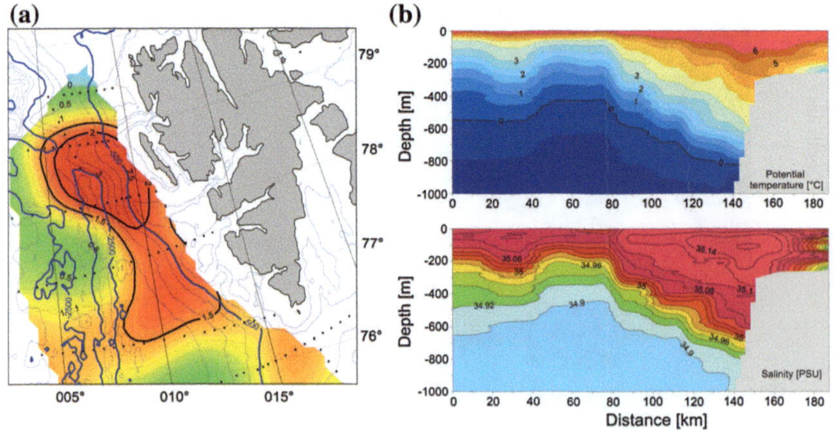

Fig. 8.6 Summer 2006. The AW heat anomaly (GJ/m^2) near Spitsbergen. 500 and 2,500 m isobath are bold. On the *right*—transect Z (temperature and salinity) along the parallel of 78°10′

between the shelf break and underwater ridge. An area of increased heat content, probably a remainder of the eddy's advection, was found south-west of the eddy. Similarly to the northern eddy in 2005, the eddy probably moved from above the Knipovich Ridge towards the Spitsbergen shelf. The same mechanism might have occurred in both cases, and in 2006 we observed a later stage of the same process. While moving northward along the ridge, the eddy reached the region where isobaths converge. Its eastern part moved over the Spitsbergen shelf break (Fig. 8.6, on the right) and the advection in the right part slowed down and the western part of the eddy started catching up with it. The eddy "rolled over", sitting on the shelf and leaving a trace of warmer water—a result of partial dissipation. The anomaly calculated for the eddy core—along the 2 GJ/m^2 isoline—covers the area of 8,000 km^2 and contains 2.5 × 10^{19} J of heat surplus. Total heat accumulated in that eddy (calculated in reference to temperature –0.1 °C) was estimated on 8 × 10^{19} J. It can sustain heat transport of 46 TW for 20 days. If compared to the mean baroclinic heat transport across transect EB of 12 TW (Chap. 7), such an eddy carries heat sufficient for 77 days of such transport. The area where the eddy left its trace, within the 1.5 GJ/m^2 isoline, covers as much as 24,000 km^2, and the heat anomaly equals 5.8 × 10^{19} J. According to the author (Walczowski and Piechura 2007) the eddy is the same structure as observed in summer 2005 at the latitude of 72°–74°N. Mean advection velocity (1.9 cm/s), heat anomaly and the structure is the same. The eddy probably reached the Arctic Ocean. While participating in hydrographic research during a cruise of German RV "Maria S. Merian" in autumn 2006, the author observed an extraordinary increase of temperature at the transect EB, along the parallel 78°50′N. It was probably the same eddy as observed in summer 2006 below the Fram Strait.

Eddies similar to described above may be of particular importance for the heat transport into the AO. Assuming that the southern eddy observed in 2005 maintained and carried to the Arctic Ocean the heat anomaly it stored, it would be capable of melting additional 1.8×10^{12} m^3 of ice, which equals 1.8×10^5 km^2 of 1 m thick ice. The assumption that the eddy may carry half of that heat anomaly seems to be reasonable, if we analyze the results from 2006. The remaining heat may affect the region by interactions with the atmosphere or advection into the AO with the WSC stream. It is obviously not possible that the entire heat has been used for melting ice, however, it shows the scale of heat transported into the Arctic by similar eddies. The heat surplus supplied by the eddy, as compared to mean transport, impacts the heat budget of the Arctic Ocean, and affects ice melting and warming of the basin. The eddy releases carried heat in pulses concentrated in a relatively small area. Polyakov et al. (2005) wrote that similar eddies may seriously affect the hydrography of the Arctic Ocean. Moorings along the Arctic Ocean shelf recorded a sudden increase of the AW layer temperature probably caused by passing of a large eddy. The origin of the eddies is not known, they are much larger than the ones previously observed and described (Walczowski 1997), forming in the Arctic Front. Such great structures probably form upstream, in the source zone of the western WSC branch or in the Lofoten Basin.

The Institute of Oceanology has been observing the relationship between the ice cover in the north-eastern part of the Fram Strait and north of Spitsbergen, and the AW temperature. According to recently prevailing opinions, the Atlantic Water flowing into the AO does not participate in ice melting, since after flowing into the AO it undergoes subduction, is covered by a layer of lighter (colder, yet less saline) Atlantic Water, is separated with cold halocline and has no contact with the atmosphere. However, the recent years show that the Atlantic Water can have a significant influence upon the ice cover not only in the Fram Strait area but also on the opposite side of the Arctic Ocean. Various branches inflowing through the Fram Strait can operate in various regions. The Svalbard Branch probably interacts with ice at a large distance from the Fram Strait, the AW flows along the shelf break, covered by a thick layer of the Polar Water. Dynamic factors, such as upwelling and interactions with the bottom, may support the AW contact with the surface. The Svalbard Branch may transport most of the heat accumulated prior to subduction along large distances. The central branch flowing into the Arctic Ocean—the Yermak Branch, flowing over the shallow Yermak Plateau interacts with the atmosphere and ice easier, thus, operating in a local range. It indicates that, depending on the intensity of the Svalbard or the Yermak Branch, ice will be melted in direct vicinity of the Fram Strait or at a great distance from it. According to the above-described original concept of the Yermak Branch origin as a part of the western branch (Chap. 6), during the 2004–2007 intensification of the western branch, ice cover near Spitsbergen should have intensely melted. The phenomenon did in fact occur—ice-free area near Spitsbergen was unprecedently large in those years. In the local scale, the Atlantic Water affects ice coverage of shelves and in fjords of the Western Spitsbergen by influencing the atmosphere and inflow of warm and highly saline AW into the fjords.

References

Cottier FR, Nilsen F, Inall ME, Gerland S, Tverberg V, Svendsen H (2007) Wintertime warming of an Arctic shelf in response to large-scale atmospheric circulation. Geophys Res Lett 34:L10607. doi:10.1029/2007GL029948

Polyakov IV et al (2005) One more step toward a warmer Arctic. Geophys Res Lett 32:L17605. doi:10.1029/2005GL023740

Walczowski W, Piechura J (2007) Pathways of the Greenland Sea warming. Geophys Res Lett 34:L10608. doi:10.1029/2007GL029974

Walczowski W (1997) Transfrontalna wymiana masy i ciepła w rejonie Frontu Arktycznego, Ph D Thesis, IOPAN, p 123

Chapter 9
Final Conclusions

The paper presents a wide description of hydrographic conditions in the studied area of the Norwegian and Greenland Seas. Also the variability of the conditions have been analysed using time series of the Atlantic Water properties measured each summer from 2000 to 2007. The warming observed in that period has been studied in detail as well as cooling of the Atlantic Water flowing towards the Fram Strait in 2007. Furthermore, the new ideas regarding multi-branch structure of the West Spitsbergen Current have been presented, including dynamical properties of flow in individual branches, as well as their variability. Based on analysis of large mesoscale structures, mean advection velocity in the western branch has been determined.

Description of the structure, variability and transport of sea currents is based mostly on hydrographic measurements and baroclinic calculations. It provides a stable image of thermohaline circulation in the studied region. There is no possibility to separate directly or distinguish between the effects of different forcings, however, the baroclinic currents provide good approximation of thermohaline circulation. The image of THC that emerges from the described series of measurements is optimistic. We are not in danger of thermohaline circulation collapsing in near future, despite the fact that such a vision of abrupt climate change resulting from a shut-down of the THC has been sometimes presented in recent years. Furthermore, it was confirmed that the baroclinic currents carry enough heat to generate ocean–atmosphere heat fluxes up to 100 W/m^2 in the AREX study area. The result acknowledges the importance of thermohaline circulation for the climate and baroclinic forces as drivers of sea currents. An additional meaningful result is finding a very good correlation between mean annual temperature of air in Hornsund and the AW temperature. That confirms a leading role of oceanic heat in climate forming.

In recent years, the West Spitsbergen Current has carried warmer AW northwards, and the increase of temperature has been compensated by the increase of salinity. The AW cooling was observed in summer 2007, which suggests that the peak of temperature maxima in the WSC has come to an end and a gradual cooling occurs. However, that does not mean that the Arctic Ocean will begin to freeze to the same degree as tens of years ago. The entire climate system of that region,

W. Walczowski, *Atlantic Water in the Nordic Seas*,
GeoPlanet: Earth and Planetary Sciences, DOI: 10.1007/978-3-319-01279-7_9,

including sea currents, is very strongly connected with global climate through different feedback mechanisms. The effects of global warming continue to affect the Arctic via the atmosphere and the ocean. The structure of wind fields changes, the air temperature rises. The growing trend in AW temperature will certainly continue, since it is mostly related to temperature variability in the subtropics. The Arctic responds to varying forcings in a different ways, including a change in the intensity of surface and deep sea currents, enhanced ocean–atmosphere exchange, ice drift and export, melting of glaciers and sea ice. The latter change, a disastrous reduction of sea ice cover seems to be the greatest, yet, very poorly recognized danger. That non-linear process, once started, will be very difficult to stop, and today we are unable to estimate its future significance, range or even time scale. The IOPAS observations indicate the relationship between the amount of heat carried by the AW and the ice cover concentration. In the present paper, the author included some suggestions on how the structure of the AW branches in the Fram Strait may affect the ice distribution. Generally, it seems that the intensification of the western and central branches affects the ice locally—near Spitsbergen, the strong Svalbard Branch impacts (with a time lag) distant areas, even in the Chukchi Sea region. The impact of the Return Atlantic Water on sea ice melting near Greenland is poorly recognized. That process may be of key importance for the Arctic Sea ice cover and depends on the transfrontal transport across the Arctic Front (Chap. 7).

Another problem emerges here—our ability to predict environmental changes. In order to predict them, they must be thoroughly recognized and comprehended. Unfortunately, that is not enough. Even if we could understand all the mechanisms, the unpredictability of forcings variability could limit our forecasting. In the present paper the author decided to restrict to studies of statistical correlations between various physical properties or the dynamics of the Atlantic Water. Unfortunately, the result does not provide a clear picture of the processes, time lags between changes or causal relationships between the processes. One can only draw general conclusions which are sometimes contradictory. The variable nature of the AW properties as well as the transport and heat exchange is more complex than it may seem, since two main forcings are at play: the wind field and the thermohaline relations. Yet, some relationships seem promising and enabled to predict (in a time scale of one year) changes in the Atlantic Water temperature. Only further measurements and gathering longer time series of data may verify whether the above concepts are correct.

9.1 The Possibilities of Diagnosis

The presented results indicate that there are some possibilities to obtain a diagnostic picture of average conditions in the entire AREX study area on the basis of individual transects. The idea that from knowledge of mean hydrographic conditions at

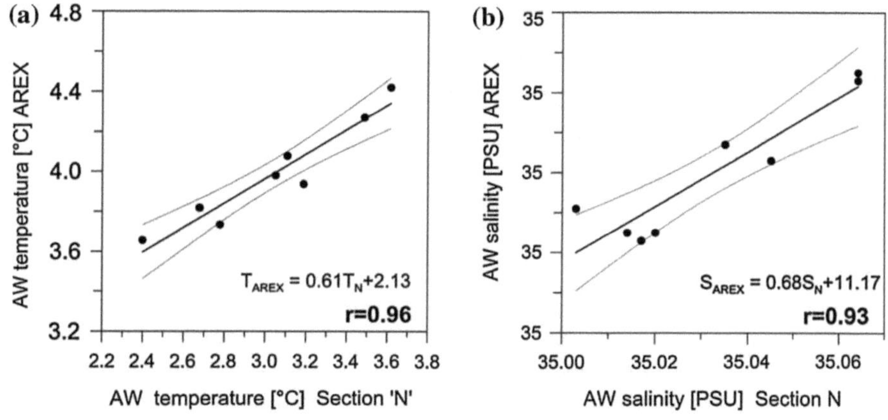

Fig. 9.1 Linear regression of AW properties (with trend) on transect N and the AW properties on the entire AREX study area. **a** temperature, **b** salinity. 95 % confidence interval is marked

one transect it is possible to predict (approximately) mean conditions at other transects, or in the entire area of study, is based on the calculations and plots shown in Sect. 5.7. A nearly linear decrease of mean temperature and salinity in the function of latitude for the period 2000–2007 is distinct. It also applies to the transect area of the AW layer and heat content. The temperature and salinity time series presented for each year separately (Fig. 5.21) also confirm a linear decrease of those parameters along latitude in each case (except for 2003). Mean decrease of temperature is ca. 0.21 °C/100 km and salinity falls by 0.009 1/100 km. That result is also confirmed by the analysis of transects. The best example is transect N described in Sects. 5.3 and 7.6. In the period 2000–2007 there is a very high correlation between the averaged AW properties from the entire AREX study area and the same properties from transect N (Fig. 9.1). For linear regression, the confidence intervals (at the level of 95 %) are narrow, which confirms that the regression lines matched the data well. Similar to Sect. 5.8, all correlation coefficients as well as the linear regression equations shown in the figures are calculated for data with trend. The correlation coefficients (usually lower) for detrended data are also provided in Sect. 5.8. The AW properties on transect 'N' also correlate best with the AW properties at other transects, measured in the same year (Sect. 5.8).

The correlations rise when comparing the AW properties at the transect N and properties in the northern part of the AREX study area (Fig. 9.2). Similar correlations for the southern part of the AREX study area are weaker: $r = 0.82$ for temperature and $r = 0.84$ for salinity.

Also the indicators of the region's dynamics (mean value of the V and V_{30} component of currents at the level of 100 dbar) correlate well with the AW volume transport across transect N. More of such relations were found, however, only the

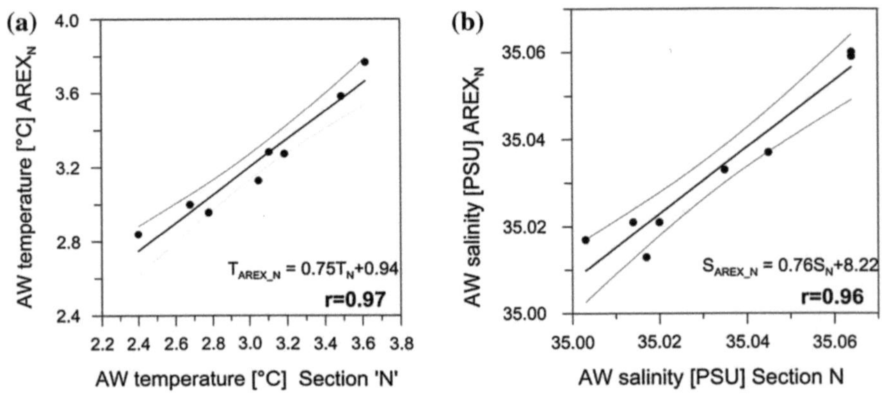

Fig. 9.2 Linear regression of AW properties (with trend) on transect N and the AW properties for the northern part of the AREX study area (above 74 N). **a** temperature, **b** salinity. 95 % confidence interval is marked

most robust and useful for predition purposes have been described herein. Obviously, that does not mean that measurements at the transect N may replace measurements in the entire study area. However, they can seriously help in reconstructing mean AW properties in the past and in forecasting future development. The important aspect is that the transect N or another section situated below Sørkapp are among the most frequently studied in that part of the Nordic Seas.

9.2 Prediction Potential

Another difficult and important issue is the potential of predicting properties and dynamics of the AW in the studied region on the basis of the existing time series or measurements performed in previous years. Very low values of the autocorrelation coefficients clearly indicate that data obtained one year are not suitable for predicting the AW properties in the same point the following year. Yet it seems obvious that if properties of the inflowing AW are acquired in the region of its formation and carried farther by advection, the upstream data may serve as a basis for predicting the AW properties in the downstream region, which is farther north in the case of the WSC. It was mentioned before that on its way the AW is modified by different processes, frequently contradictory in effect. Thus, the prediction potential based on the *upstream* data is very limited. Nevertheless, some useful relationships have been found.

There is a high correlation between salinity at the transect H and salinity at transect N the following year (Fig. 9.3). For temperature the relation is much

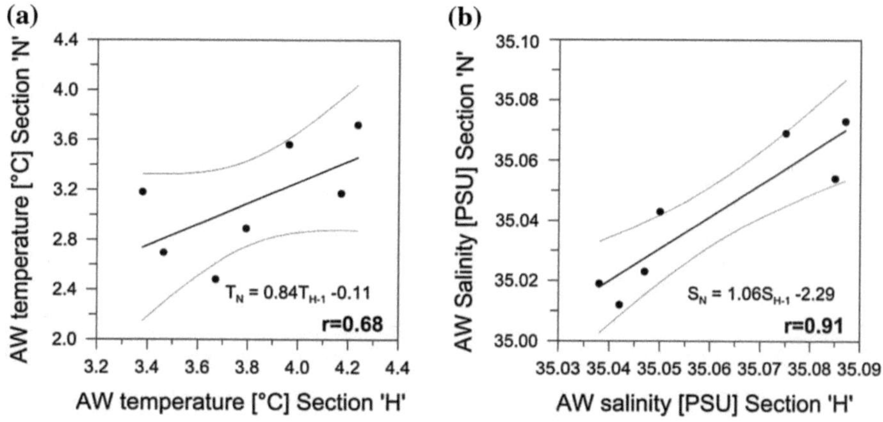

Fig. 9.3 Linear regression of AW properties (with trend) at the transect H and the AW properties at the transect N the following year. **a** Temperature, **b** salinity. 95 % confidence interval is marked

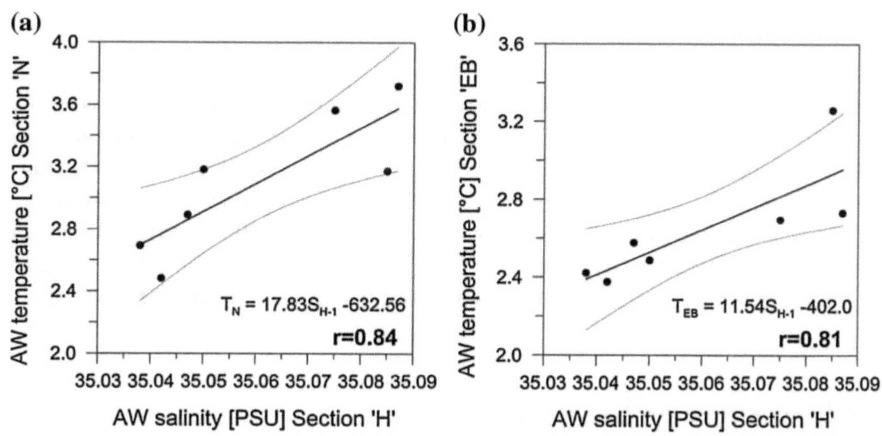

Fig. 9.4 Linear regression of AW salinity (with trend) at the transect H and the AW temperature at northern transects the following year. **a** transect N, **b** transect EB. 95 % confidence interval is marked

weaker, the correlation coefficient lower, and confidence intervals wider. The temperature trend at northern transects may be, however, successfully predicted on the basis of salinity at the southern transect (H) in the previous year (Fig. 9.4).

The suitability of salinity values for predicting changes in the following year is also confirmed by the correlations calculated between mean values from the northern study area and the southern study area (Fig. 9.5). Section 4.5 presented the detailed analysis of more correlations.

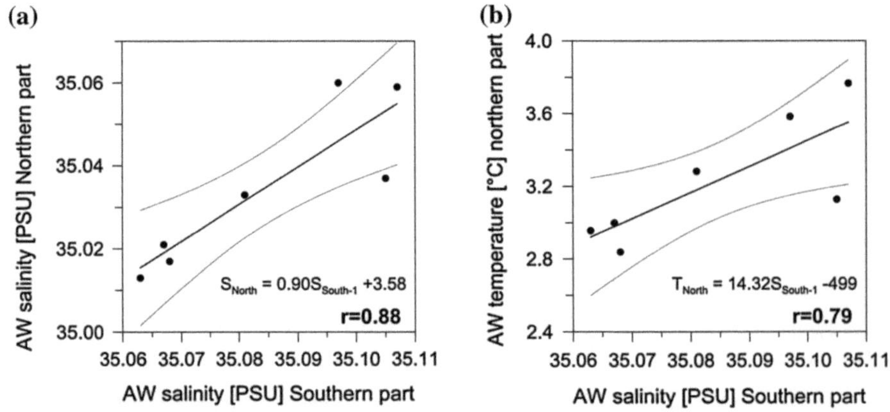

Fig. 9.5 Linear regression of AW salinity (with trend) in the southern part of the AREX study area (**a**), salinity and temperature in the northern part of the study area the following year (**b**). 95 % confidence intervals are marked

9.3 Summary

A great deal of meaningful and some less important conclusions may be drawn from the extensive collection of observations. The author is aware that he has used only a part of the information contained in this data collection and hopes that it will be utilized in future. The present paper could also support future studies. In the last section, the conclusions drawn from the paper will be revisited and presented in a systematic way. Some of them are obvious, whereas others are quite original. However, they may not be separated not to lose the perspective view.

9.3.1 Structure of Currents

- The heat delivered into the Arctic Ocean through the Fram Strait by the Atlantic Water originates from the advection of that water from the south and it was previously accumulated at lower latitudes;
- The inflow of the AW into the Nordic Seas is split into three main branches: the North Iceland Current, the Faroe Current and the Shetland Current;
- In the southern area of the studied region two AW branches are observed: a continuation of the Faroe branch and the Shetland branch;
- The Atlantic Water in those branches was formed in various regions of the north Atlantic and has different physical properties. The AW in the western branch is colder and less saline than the one in the eastern branch;
- The structure and the dynamics of flow are different in both branches;

- The eastern branch (the WSC core) flows into the Norwegian Sea between the Faroe and the Shetland Islands and continues along the shelf edge of Norway, the Barents Sea and the western Spitsbergen mostly as a barotropic flow, and later as a barotropic-baroclinic flow;
- The western branch of the AW is a continuation of the Faroe branch which flows above the Iceland-Faroe Ridge. In the area of study the branch flows on the eastern side of a series of underwater ridges (the Mohn Ridge, the Knipovich Ridge) and is mostly baroclinic;
- At the latitude of 74°N the western branch may split, the inner part of the branch may flow into the central part of the Atlantic Domain and join the eastern branch;
- Due to the bottom topography, all the WSC branches approach or even converge near the latitude of 78°N.
- The region of converging isobaths, at the latitude of approximately 78°N, plays a role of a gateway for the AW inflow into the Arctic Ocean;
- North of the narrowing the current divides again into two or three branches: the Svalbard Branch flowing along the shelf break eastward into the AO, the Yermak Branch flowing into the AO northward over the Yermak Plateau, and the western branch recirculating towards the south-west and the south;
- The occurrence and intensity of the central Yermak Branch probably depends on the intensity of the western branch and its splitting at the latitude of 74°N;
- The velocity of signal propagation in the eastern branch was estimated on 3 cm/s on the basis of the IOPAS measurements at the transect 76°30'N and the data available from the Svinoy Section;
- Observing the same eddy for two following years allowed to estimate mean advection velocity in the western branch on 1.9 cm/s;
- The baroclinic transport across the BSO is close to the value determined on the basis of measurements performed with a moored array. That is odd, considering that fact that the WSC is assumed to be strongly barotropic;
- The baroclinic volume and heat transports across the northernmost transect are much lower than the transport determined from measurements with moorings in the Fram Strait.

9.3.2 Spatial Changes

- The area of study is naturally divided into 2 parts: the southern one and the northern one;
- Due to different properties of the currents, it can be also divided into the eastern and the western part;
- The dynamics of the southern part is related mostly to the inflow of the AW from the south and the south-west and the AW transport into the Barents Sea;
- In the northern part the AW is transported towards the Fram Strait; also intensive AW transport across the Arctic Front occurs;

- During advection into the AO the AW undergoes strong transformation by processes such as exchange of heat and moisture with the atmosphere, mixing with the surrounding waters, convergence and mixing of the various current branches, divergence of fluxes;
- The above processes (particularly mixing with the surrounding waters) intensify at the latitude of the Barents Sea Opening;
- As a result of the transformation, the range of variability of the AW properties (standard deviation) is larger in the northern region than in the source region;
- A linear decrease of temperature and salinity in the AW layer was observed with increasing latitude;
- Mean area of the AW section perpendicular to the flow also decreases linearly as the latitude increases;
- North of the 75°N parallel, the mean baroclinic flow also decreases linearly as the latitude while it is not the case south of 75°N, probably due to the AW circulation in the Bear Island Trough;

9.3.3 Temporal Variability

- The temporal variability of properties of the AW flowing into the Nordic Seas is associated mainly with the variability of meteorological and hydrographic conditions in the area of water mass formation;
- In the southern part of the studied area, the hydrographic properties of the AW in the eastern branch are steadier than in the western branch;
- A sudden decrease of the AW temperature was observed in the period 2000–2003 in the studied area, and from 2004 to 2006 there was a sudden increase of temperature and heat content in the AW layer;
- In the 2004–2006 period the AW significantly expanded northward and the ocean climate in the Fram Strait area changed. The summer position of the 5 °C isotherm at 100 m shifted 300 km north;
- The AW temperature fell significantly in 2007;
- The salinity reacted in a similar way as temperature, sometimes (2006) with an advance of one year;
- The increase of heat content in the AW layer was associated with the increase in temperature, the AW layer thickness decreased during that time;
- The western branch significantly contributed to the 2004–2006 warming, it was much stronger than before;
- The baroclinic transport towards the Barents Sea prevailed until 2004, after 2004 the transport towards the Fram Strait intensified;
- Large mesoscale eddies were observed in 2005 and 2006 which presumably played an important role in the AW warming observed in the AREX study area,

as well as in warming of the West Spitsbergen fjords and heat transport into the AO;
- The origin of the eddies is not known, they probably formed south of the study area;
- There is a coincidence between the heat content in the AW layer and the sea ice cover in the area north and north-east of Svalbard. With greater heat content and higher AW temperature the ice-free area is larger;
- The decrease in the AW layer temperature and heat content in 2007 in the northern part could be predicted from observations in the southern part of the AREX area in 2006;

9.3.4 Correlation Relationships

- Transport of properties acquired in the area of the Atlantic Water formation suggests that by observing the variability at lower latitudes it is possible to predict the AW properties at higher latitudes;
- Transformation of the AW properties results in modifications of original properties, therefore, strong temperature or salinity anomalies are the most suitable for studying signal propagation;
- In spite of the transformation, meaningful correlations were found between time series of the AW properties at individual transects;
- The correlation coefficient decreases with distance, however, correlation is often higher between transects situated at a medium distance from each other (300 km) than between those which are close or far away from each other;
- The autocorrelation with 1 year's time lag is always very low. It means that on the basis of measurements at a given transect one cannot predict the AW properties at that transect the following year;
- However, some correlations between the AW properties at the southern transects (or in that region) and the AW properties at the northern transects (or in that region) the following year were observed;
- The AW salinity in the southern part is a good proxy for predicting changes in the following year in the northern part;
- Relations between the AW physical properties and the sea region dynamics were found;
- AW temperature is positively correlated with the intensity of northward flow;
- The increase of baroclinic transport towards the Barents Sea is accompanied by the decrease of northward transport;
- The increase of baroclinic transport towards the Barents Sea is accompanied by the decrease of the AW temperature in the northern part the following year;
- Only weak negative correlations between the AW properties and the winter NAO index were found, maximum values are reached for 3–4 years' time lag;

9.3.5 The Ocean Climate and the Significance of the AW for the Climate

- A very high correlation between mean annual air temperature in Hornsund and the AW temperature was found;
- It confirms the climatic significance of the AW inflow for the region;
- It was shown that the heat supplied by baroclinic currents alone is capable of supporting large ocean–atmosphere heat fluxes;
- The relationship between ice cover around Spitsbergen and the AW temperature was found;
- The ice cover north-east of Spitsbergen can be influenced by the structure of currents in the Fram Strait;
- Large AW transport across the Arctic Front may have an indirect impact upon the AO ice cover;
- A significant change of the ocean climate was observed in the period 2005–2006, caused mainly by the increase of the currents velocities and intensification of the western WSC branch;
- An important role is played by baroclinic eddies of unusually large diameters observed in the western branch;
- The changes affected the hydrography and ecosystem of the Spitsbergen fjords as well.

The data gathered by the IOPAS and analyzed herein are not sufficient to answer some general questions regarding the causes or effects of global warming. Based on the data we have, it is impossible to state whether the observed AW warming is of the first (a change of global heat balance), or the second (a change of heat redistribution) type. On the other hand, the presented results provide a new knowledge on the possible paths and mechanisms of the AW warming in the studied area, the AW transformation, variability of heat transport, and, finally, on the role of the ocean heat transported by the Atlantic Water into the Nordic Seas. They also show that the thermohaline circulation in that region did not weaken as a result of increased transport of fresh water into the system.

Commencing work in the area of a great importance for the thermohaline circulation makes one aware how much there is still to do in the field of the physical oceanography. The author hopes that the work will improve the knowledge on the studied area and help develop a strategy for future studies.

Bibliography

Aagaard K, Greisman P (1975) Toward new mass and heat budgets for the Arctic ocean. J Geophys Res 80:3821–3827

Aagaard K, Swift JH, Carmack EC (1985) Thermohaline circulation in the Arctic Mediterranean seas. J Geophys Res 90:4833–4846

Andersen C, Jennings A, Andrews IJT (2004) Non-uniform response of the major surface currents in the Nordic seas to insolation forcing: implications for the Holocene climate variability. Paleoceanography 19:PA2003. doi:10.1029/2002PA000873

Bryden HL, Imawaki S (2001) Ocean heat transport. In: Siedler G, Church J, Gould J (eds) Ocean circulation and climate. Academic Press, Waltham, pp 455–474

Bryden HL, Longworth HR, Cunningham SA (2005) Slowing of the Atlantic meridional overturning circulation at 25°N. Nature 438(1):655–657. doi:10.1038/nature04385

Dera J (2003) Fizyka Morza. Państwowe Wydawnictwo Naukowe, Warszawa, p 541

Druet C (2000) Dynamika Morza. Wydawnictwo Uniwersytetu Gdańskiego, Gdańsk, p 281

Druet C, Jankowski A (1991) Flow across south and east boundaries of the Norwegian sea. Oceanologia 30:37–46

Druet C, Jankowski A (1992) Some results of three—year investigations on the interannual variability of the Norwegian—Barents confluence zone. Pol Polar Res 13(1):3–17

Fischer J, Visbeck M (1993) Deep velocity profiling with self-contained ADCPs. J Atmos Oceanic Technol 10(5):764–773

Furevik T (2001) Annual and interannual variability of Atlantic water temperatures in the Norwegian and Barents seas: 1980–1996. Deep-Sea Res I 48:383–404

Gregory JM, Weaver AJ, Driesschaert E, Eby M, Fichefet T, Hasumi H, Hu A, Jungclaus JH, Kamenkovich IV, Levermann A, Montoya M, Murakami S, Nawrath S, Oka A, Sokolov AP, Thorpe RB (2005) A model intercomparison of changes in the Atlantic thermohaline circulation in response to increasing atmospheric CO_2 concentration. Geophys Res Lett 32:L12703. doi:10.1029/ 2005GL023209

Hansen J, Nazarenko L, Ruedy R, Sato M, Willis J, Genio DA, Koch D, Lacis A, Lo K, Menon S, Novakov T, Perlwitz J, Russell G, Schmidt G, Tausnev N (2005) Earth's energy imbalance: confirmation and implications. Science 308(5727):1431–1435

Hátún H, Sandø AB, Drange H, Hansen B, Valdimarsson H (2005) Influence of the Atlantic subpolar gyre on the thermohaline circulation. Science 309:1841–1844

ICES (2005) The annual ICES ocean climate status summary 2004/2005. ICES cooperative research report no. 275, p 37

Johannessen OM, Bengtsson L, Miles M, Kuzmina SI, Semenov V, Alekseev GV, Nagurny AP, Zakharov VF, Bobylev LB, Petterson LH, Hasselman K, Cattle H (2004) Arctic climate change: observed and modelled temperature and sea ice. Tellus Ser A 56:328–341

Levitus S, Antonov JI, Boyer TP (2005) Warming of the world ocean, 1955–2003. Geophys Res Lett 32:L02604. doi:10.1029/ 2004GL021592

W. Walczowski, *Atlantic Water in the Nordic Seas*,
GeoPlanet: Earth and Planetary Sciences, DOI: 10.1007/978-3-319-01279-7,

Macrander A, Send U, Valdimarsson H, Jonsson S, Kaese RH (2005) Interannual changes in the overflow from the Nordic seas into the Atlantic ocean through Denmark strait. Geophys Res Lett 32:L06606. doi:10.1029/2004GL021463

Maslowski W, Marble DC, Walczowski W, Semtner AJ (2001) On large scale shifts in the Arctic ocean and sea ice conditions during 1979–1998. Ann Glaciol 33:545–550

Orvik KA, Skagseth Ø, Mork M (2001) Atlantic inflow to the Nordic seas: current structure and volume fluxes from moored current meters. VM-ADCP and seasoar-CTD observations, 1995–1999. Deep Sea Res 48(4):937–957

Piechura J (1996) Dense bottom waters in Storfjord and Storfjordrenna. Oceanologia 38(2):285–292

Piechura J, Osiński R, Walczowski W (2000) Some results of exchanges through the Norway-Svalbard opening. News Lett Eur Geophy Soc 74:1645

Poulain PM, Warn-Varnas A, Niiler PP (1996) Near-surface circulation of the Nordic seas as measured by Lagrangian drifters. J Geophys Res 101(C8):18237–18258

Przybylak R (1966) Zmienność temperatury powietrza i opadów atmosferycznych w okresie obserwacji instrumentalnych w Arktyce. Uniwersytet Mikołaja Kopernika w Toruniu, Toruń, p 279

Rahmstorf S (1999) Shifting seas in the greenhouse? Nature 399:523

Rudels B (1987) On the mass balance of the Polar ocean, with special emphasis on the Fram Strait. Norwegian Polar Institute, Skr.188, p 53

Rudels B, Bjork G, Nilsson J, Winsor P, Lake I, Nohr C (2005) The interaction between waters from the Arctic ocean and the Nordic seas north of Fram Strait and along the east greenland current: results from the Arctic ocean-02 Oden expedition. J Mar Syst 55:1–30

Schauer U, Muench RD, Rudels B, Timokhov L (1997) The impact of eastern Arctic shelf waters on the Nansen basin intermediate layers. J Geophys Res 102:3371–3382

Skagseth Ø (2004) Monthly to annual variability of the Norwegian Atlantic slope current: connection between the northern north Atlantic and the Norwegian sea. Deep-Sea Res I 51:349–366

Steele M, Morley R, Ermold W (2003) PHC: a global PHC 3.0, updated from: ocean hydrography with a high quality Arctic ocean. J Clim 14(9):2079–2087

Stewart R (2006) Introduction to physical oceanography. Texas A&M University, Texas, p 343

Styszyńska A (2005) Przyczyny i mechanizmy współczesnego ocieplenia 1982–2002 atlantyckiej Arktyki. Wydawnictwo Uczelniane Akademii Morskiej w Gdyni, p 109

Walin G, Broström G, Nilsson J, Dahl O (2004) Baroclinic boundary currents with downstream decreasing buoyancy; a study of an idealized Nordic seas system. J Mar Res 62(4):517–543

MIX
Papier aus verantwortungsvollen Quellen
Paper from responsible sources
FSC® C105338

If you have any concerns about our products,
you can contact us on
ProductSafety@springernature.com

In case Publisher is established outside the EU,
the EU authorized representative is:
Springer Nature Customer Service Center GmbH
Europaplatz 3, 69115 Heidelberg, Germany

Printed by Libri Plureos GmbH
in Hamburg, Germany